Praise for *Liars, Lovers, and Heroes*

"Research in molecular biology, cognitive neuroscience, and artificial intelligence is radically changing the view humans have of themselves and society. In *Liars, Lovers, and Heroes,* Steven Quartz and Terrence Sejnowski bring their expertise in theoretical biology and philosophy to bear on the new results and reflect on their meaning. Their book provides a rare opportunity to be informed intelligently about the changing world of modern human science."   —Antonio R. Damasio, Van Allen Professor of Neurology, University of Iowa College of Medicine; author of *Descartes' Error: Emotion, Reason, and the Human Brain*

"A superb book. . . . A breath of fresh air."   —V. S. Ramachandran, M.D., Ph.D., professor and director, Center for Brain and Cognition, University of California, San Diego; adjunct professor, Salk Institute; author of *Phantoms in the Brain*

"An evocative solution to a classic problem: Which is more important in shaping the human brain, nature or nurture?"   —Sandra Blakeslee, *New York Times*

"Wide-ranging . . . linking cutting-edge neuroscience with social history and popular culture . . . [with] postmodern culture and globalization."   —*Publishers Weekly*

"Neuroscience seems prone to coming up with polarizing theories of personality: you are either your genes or your environment. Countering the standard dichotomy, this fresh approach to conceptualizing brain development from a pair of California-based researchers touts 'cultural biology.' The authors define the meaning of that term while addressing topics such as emotion, sex, and happiness—but Quartz and Sejnowski improve on those themes by informing readers how brain anatomy and neurochemistry work in focusing one's desire. Although the authors discuss serotonin, dopamine, and a reptilian vestige called the ventral basal ganglia, their text is not a clinical parade of jargon, and they are adept at using anecdotes to illustrate their points (such as why motivator Tony Robbins is optimistic and filmmaker Woody Allen is pessimistic). In accessible, conversational language, the authors offer an intriguing investigation of personality."   —*Booklist*

"Smart authors with a lot of hot stuff to report on."   —*Kirkus Reviews*

"An entertaining and startling survey of what it means to be human."

—*Discover* magazine

About the Authors

Garry Hunter

STEVEN R. QUARTZ, PH.D., is director of the Social Cognitive Neuroscience Laboratory at the California Institute of Technology, and an associate professor in the Division of Humanities and Social Sciences and the Computation and Neural Systems Program. He was a fellow of the Sloan Center for Theoretical Neurobiology at the Salk Institute and a recipient of the National Science Foundation's CAREER award, its most prestigious award for young faculty. He lives in Topanga, California.

Beatrice Golomb

TERRENCE J. SEJNOWSKI, PH.D., is regarded as the world's foremost theoretical brain scientist. His demonstration of NETtalk, a neural network that learned to read English words, helped spark the 1980s neural network revolution for which he received the IEEE Neural Network Pioneer Award in 2002. He received his Ph.D. in physics from Princeton University before studying neurobiology at Harvard University School of Medicine. He is an investigator with the Howard Hughes Medical Institute and directs the Computational Neurobiology Laboratory at the Salk Institute. At the University of California at San Diego, he is a professor of biology, physics, and neuroscience, and directs the Institute for Neural Computation. He has published more than two hundred scientific articles, and has been featured in the national media. He lives in Solana Beach, California.

liars, lovers, and heroes

WHAT THE NEW BRAIN SCIENCE REVEALS ABOUT HOW WE BECOME WHO WE ARE

Steven R. Quartz, Ph.D.,
and Terrence J. Sejnowski, Ph.D.

Quill
An *Imprint of* HarperCollins*Publishers*

A hardcover edition of this book was published in 2002 by William Morrow, an imprint of HarperCollins Publishers.

HarperCollins books may be purchased for educational, business, or sales promotional use. For information please write: Special Markets Department, HarperCollins Publishers Inc., 10 East 53rd Street, New York, NY 10022.

First Quill edition published 2003.

Designed by Nicola Ferguson

The Library of Congress has catalogued the hardcover edition as follows:

Quartz, Steven.
Liars, lovers, and heroes : what the new brain science reveals
about how we become who we are / Steven Quartz and Terrence Sejnowski.
p.   cm.
Includes bibliographical references and index.
ISBN 0-688-16218-5
1. Genetic psychology.   2. Brain—Evolution.   I. Sejnowski,
Terrence J. (Terrence Joseph).   II. Title.
BF701.Q83 2002
612.8'2—dc21   2001026636

ISBN 0-06-000149-6 (pbk.)

    05 06 07 ❖/RRD 10 9 8 7 6 5 4 3 2

*For Francis, whose scientific pursuit of big
questions has inspired us and whose generous
advice has helped guide us*

contents

acknowledgments

This book owes its existence to the lively conversations that began every day at 3:30 P.M. with tea and treats in the Computational Neurobiology Laboratory at the Salk Institute for Biological Studies. The lab has been an extraordinary source of new ideas and stimulating discussion over the last twelve years. We have benefited from the many students, friends, and colleagues with whom we have interacted in the CNL tearoom as well as all who have visited. Rosemary Miller has kept the good ship CNL afloat for generations of talented modelers.

We have been privileged to interact with a number of extraordinary groups: Members of the La Jolla Initiative for Explaining the Origins of Humans provided numerous insights into many of the central issues we explore in this book. Conversations with members and guests of the Helmholtz Club also provided many insights into the brain.

We have a special debt to those who read and commented on earlier drafts: John Allman of Caltech; Beatrice Golomb, James Moore, and John Kelsoe of the University of California at San Diego and Read Montague at Baylor College of Medicine; and Francis Crick of the Salk Institute. We are deeply grateful to our editor, Toni Sciarra, who in reading drafts seemed able to read our minds and thereby translated our intentions into prose. Katinka Matson and John Brockman patiently helped this project along and provided invaluable advice.

Steve would like to acknowledge his gratitude to Patricia Churchland and Terry Sejnowski, coadvisers who always encouraged him to follow problems rather than disciplinary boundaries. Terry's lab was an ideal place to learn about the brain. There, Richard Adams provided patient counsel on confocal microscopy while Shona Chattarji revealed the physiological life of neurons, and Read Montague, Alex Pouget, Peter Dayan, and many others were always game to discuss both the large and the small. The good fortune of being surrounded by outstanding colleagues continues at Caltech, where Jim Woodward, Fiona Cowie, Dominic Murphy, Alan Hajek, Chris Hitchcock, Christof Koch, John Allman, Shin Shimojo, Colin Camerer, and others create an extraordinary intellectual climate. Steve also feels fortunate to belong to the wonderful group Kathleen Akins has brought together in the McDonnell Project in Philosophy and the Neurosciences. Michael Dobry and the students in the graduate industrial design program at the Art Center College of Design in Pasadena provided a welcoming environment to discuss these ideas and to think about the relationship between the nervous system and the world we design.

Steve wouldn't even have thought of undertaking this project were it not for Karen. The journey together that began fifteen years ago makes the questions this book explores worth asking. Along the way, Evelyn, Alden, and Elliot gave new meaning to the answers. They have spent much of their childhood curious as to why adult books take so long to finish, but their faith that it would be completed one day provided a source of encouragement.

Terry's education in neuroscience began at Princeton, with exciting discussions with Mark Konishi, Charles Gross, Alan Gelperin, and especially John Hopfield, who made possible his passage from physics to biology. Steve Kuffler at Harvard Medical School introduced Terry to the mysteries of neurons and showed him the dignity of the single synapse. More recently, Terry's collaboration and friendship with Mircea Steriade has been a source of inspiration and revelation into the brain's secret nightlife.

Geoffrey Hinton has been an especially important influence on a generation of neural network modelers, and Terry is fortunate to have been his collaborator and friend over the last thirty years. Charles Rosenberg and Sidney Lehky made major contributions to Terry's laboratory at the Johns Hopkins University that opened up new directions for modeling brain systems. Patricia Churchland was a wonderful guide to neurophilosophy and taught him what it was to write a book.

Terry is indebted to Beatrice Golomb, who anticipated many of the ideas that have emerged from the neural network revolution. As a medical researcher she has pointed out that there is something that is even more complex than the brain—namely the body, of which the brain is a part. The mysteries of the body are vast and medicine is only beginning to understand them. Without a body there would also not be a mind-body problem.

The Salk Institute has been a wonderful research environment, and friends and colleagues at the Salk, the University of California at San Diego, and the California Institute of Technology have been a great source of wisdom in science and life, especially Francis Crick, Walter Heiligenberg, Ted Bullock, Carver Mead, and John Allman.

preface

Who are we? What is it to be a person, to love and to hate? Are we good or fundamentally evil? What makes us happy? Ultimately, who can we become?

These are some of the questions we explore in this book. Although we do not yet have the complete answers, remarkable progress in brain science now provides us with the tools that we believe are crucial to uncovering the mystery of who we are. We can now eavesdrop on the brain's swirling activity with new methods ranging from arrays of tiny electrodes to brain-imagers that scan the activity of the living human brain. Although the human brain's computational power far exceeds that of the fastest computers ever built by humans, computers are getting exponentially faster and can be used to simulate the brain. Finally, molecular genetics is providing powerful tools for manipulating brain mechanisms. Bringing the fruits of these powerful new tools together in this book, we take you on a journey inside the brain to begin to uncover the mystery of who we are.

This journey has been bolstered immensely by the culmination of the largest biological undertaking in history, the sequencing of the human genome. This critical new piece of the puzzle was just the beginning of an even more exciting search for our origins, for encoded in the human genome is not only the description of the instructions that create a human but also tantalizing hints of how humans evolved. With the immense power of genetic techniques at our disposal, we are on the verge of removing the last molecular veils behind which neurons have been able to hide. However, uncovering how brains work will require more than a complete sequence of the human genome, because—as we will explore in this book—the development of the brain and of behavior also depends on interactions with human culture. The partnership between your genes and the changing, uncertain world of culture is thus at the heart of who you are. Extracting the implications of this extraordinarily rich and complex interaction, we will introduce you to a new view of who we are that we call "cultural biology." We will explain cultural biology's emerging lessons for the enduring questions humans have long asked about what it means to be human. The themes we explore in this book were put into tragic focus by the events of September 11, 2001. In an afterword, we reflect on the main themes of this book in the light of those events and the war on terrorism.

Progress in science is made by focused experiments under highly controlled conditions, usually communicated in brief articles to scientific peers. As powerful an engine of knowledge creation as this enterprise has been, there is also value in occasionally stepping back and attempting to make connections across disciplinary boundaries. Because progress in brain and cognitive science has been so rapid in recent years, we believed such an undertaking was timely not only to make connections but also to see whether new conversations across disciplines could be started. The sciences and the humanities are both engaged in understanding who we are, and ultimately it will take a coordinated effort between them to answer this question. This book is thus an invitation to join the conversation.

In writing this book, we found it tremendously exciting to see how ideas from a wide range of fields might fit together. We realized that it also held potential pitfalls. Not only were we integrating what we knew within our own areas of research, but we were also speculating outside our expertise. We have greatly benefited from discussions with many colleagues. In some instances, the scope of our subject forced us to limit inclusion of complete scientific sources and details. A certain amount of simplification and omission were necessary as part of gearing the book to a wide audience. Any resulting inaccuracies are unintentional. For those interested in following up on these details, we have created a website for the book at www.liarsloversandheroes.com, which also includes further pointers to the scientific literature. This will be kept up-to-date as further discoveries are made and our own views on these issues evolve.

Our Brains, 1 Ourselves

The point of view taken here is neither a "biological" one nor a "sociological" one if that would mean separating these two aspects from each other.
—*Erich Fromm*, The Sane Society

There's something disturbing about holding a human brain in your hands. It feels almost sacrilegious, as though you're violating some basic taboo. Even among medical students who think nothing of eating lunch beside the cadaver they're dissecting, you can feel their uneasiness when it comes time to remove the brain. Over the electric whir of a bone saw penetrating the skull just above the ears, someone inevitably tries to ease the strain with a joke.

Cutting away the tangle of fibers holding the brain in place is slow, delicate work, increasing the tension in the room. And when the brain is finally eased out, a sickening, sucking sound breaks the silence as air rushes in to fill the void left behind.

Holding this person's brain, turning it in your hands,

you try to block the clichéd image of Hamlet pondering Yorick's skull, but it still floods your mind. You find yourself asking similar questions: What did this brain once think; what secret dreams did it hold; what memories are locked forever deep in its folds? Staring at it, you also can't blot out the thought that you are bound for a similar fate. One day all your hopes and thoughts—all that you are—will be reduced to a three-pound clump of silent cells that was once a person.

This book is about the first insights a new kind of brain science is giving us into what makes us who we are. Somehow, your sense of being someone unique—a thinking, feeling person—emerges from the workings of the most complex object in the known universe: your brain. If someone close to you has ever suffered a brain injury or disease, you know that the connection between the brain and who you are is all too real. Consider Irene's story.[1]

She had been denying the signs for a year. Then one day she had trouble finding her way home from her office, her daily commute for the last twenty years. Finally arriving home, shaken and confused, she couldn't deny it any longer. Something was happening to her, and it terrified her. Her husband, Bill, had suspected something was wrong a few weeks earlier when they'd gone out for dinner because Irene said she had forgotten to make it. When they returned home, he had found the dinner she had prepared waiting for them at the table. Trying to protect her pride, he quickly cleared it away without her seeing. But after this latest incident they both realized they couldn't deny the signs anymore.

Many doctors and tests later, Irene's worst fears were confirmed: Alzheimer's disease. Her doctor tried to explain what was probably happening in her brain. Something about plaques, tangles, and dying brain cells. But he couldn't offer much hope for treatment. As Irene's condition worsened, she would have the occasional clear day, only to break down in tears at the realization of what was happening to her. She was losing control to a disease that was attacking more than her body. It was slowly stealing her away from herself. Bill recounts how hard it was to sit by her bedside day after day when she no longer recognized him or even knew her own name, helpless to fight a disease that slowly robbed her of who she was until she was no more.

As Irene's tragedy makes painfully clear, who you are hangs in a delicate balancing act inside your brain. And although its breakdown brings unspeakable horrors, your strongest feelings of elation and even love likewise spring from your brain's intricate workings. The chemical soup inside your head gives rise to your deepest feelings, to your capacity to wonder, even to the grief that makes you human. Not all brain disorders bring the sort of tragedy that afflicted Irene. Some alter the texture our brains give to our everyday life, even making some patients curiously appreciative of the changes brought on by a stroke. Consider the strange case of a Swiss political journalist.[2] At the age of forty-eight, Klaus suffered a stroke, a rupture in the blood supply to his brain. But as he recovered in the hospital he seemed little concerned with his condition. Instead, he fixated on the hospital food, which he complained about bitterly. Now you might not think too much of that, since hospital food is pretty miserable stuff. That's what his doctor, Marianne Regard, thought at the time, too. She asked Klaus to keep a diary, and it revealed a strange change in him. Before his stroke, Klaus had always been one who ate to live. After his stroke, he lived to eat. This wasn't just bingeing on Big Macs, though. He didn't even gain weight. Instead, he had a newfound preoccupation, bordering on an obsession, with gourmet food. Damage to his brain had transformed the experience of gourmet food into one of pure rapture. Klaus even quit his job as a political journalist to become a fine-dining columnist. Since Klaus's case eight years ago, Dr. Regard has found many more cases of what's now known as "gourmand syndrome." Injury somewhere in the right frontal part of the brain can turn some people into fine-food lovers of the highest order.

As this strange brain condition suggests, every nuance of yourself, the very fabric of your experience, ultimately arises from the machinations of your brain. The brain houses your humanity. Without it, you would be as heartless and cold as a robot, an automaton for which life might have goals but would have no meaning.

Among the astounding diversity of brains in nature, ours seems capable of many unique capacities.[3] As far as we know, our brains are the only ones to generate a sense of who we are. Our brains create a life history by weaving together the events of yesterday and pondering what will become of us tomorrow. Because we possess this rare—and perhaps unique—brain capacity, we alone ask the most human of questions:

> Who are we? What makes us the way we are? What is it to be a person; to love and hate; to think and feel? Are we good or fundamentally evil; peaceful or warlike? Where did we come from? Why do we live together? What makes us happy? Do we have any power over the forces shaping us to become who we want to be? Ultimately, who can we become?

These are among the oldest questions asked. And no human group has ever existed that didn't ask these questions. Asking, probing, and searching for who you are is what makes you human. We seem driven by a basic biological need to live in a world that makes sense, whose events occur not willy-nilly but for reasons. Nowhere is the human need to make sense of experience more striking than in the case of amnesiac patients. It's almost impossible to imagine what it must be like to live the life of an amnesiac. If your closest friend or even a spouse left your room for but a minute, when they came back it would be as though you were meeting them for the very first time. But what's so striking about many amnesiacs is their readiness to confabulate, to answer questions with a shaggy dog story. In fact, when they're quizzed about themselves or even asked what day of the week it is, they often sound like fast-talking con artists selling some scheme. It's hard to suppose that someone without a past is trying to save face. They won't even remember being embarrassed about it a minute later. The often bizarre confabulations of amnesiacs suggest that our need to make sense of our world is a primal one that remains intact even when our sense of self is erased.

Perhaps this is why unexplained events are so disturbing. Why does one person's life end in a fatal highway accident, while another wins the lottery? Nearly all of us are tempted to cast these sorts of random events as part of a larger scheme, a destiny that is part of some overall coherent plan. No possibility is too fantastic if it makes life's events coherent—even alien conspiracy theories like the infamous Area 51 in Nevada. The truth is out there, we think—somewhere.

Who Do We Think We Are?

Ever since our ancestors first asked who they were, they searched for answers by looking up to the skies, crafting epic tales of creation, of fallen heroes trying to emulate the gods they worship, and of titanic struggles between good and evil. For our ancestors, fathoming human nature was a journey into a spiritual netherworld in which they imagined themselves inhabited by various life forces, perhaps a soul, that drove them to act the way they did. Then, as Darwin's revolutionary theory sank into public consciousness, we saw ourselves as just another animal, each pitted against the other in a struggle for survival. This image swept society and continues to color how we relate to others, how we raise our children, even how society and its institutions are organized. Since we are only selfish brutes incapable of solving problems of common concern on our own, the modern state itself became a grand mediator to protect us from one another.[4]

In this book, we show you the surprising answers a new kind of brain science is beginning to offer to these enduring human questions. Since these new answers need to be seen against the background of the modern image that began with Darwin, let's ponder the modern image and, in particular, how it ignited a battle over human nature that touched every facet of life in the twentieth century and continues to rage to this day.

Prison Life

The prison riot was entering its second day. The inmates had ransacked their cells, throwing everything that wasn't nailed down and smashing everything else. Barricaded, and with their paranoia growing by the minute, they waited now, wondering what was happening on the other side of the door. For the guards it was all too much. They attacked. Blasting fire extinguishers to clear the way, the guards quickly overpowered their prisoners. An uneasy quiet returned to the prison. But the guards wouldn't forget. It was payback time. And so began a campaign of retribution, with guards forcing prisoner after prisoner to perform humiliating tasks, catching them in surprise interrogations, showing them who was boss by breaking their spirit.

Just another day in one of America's maximum-security prisons? Not exactly. This prison riot didn't take place in a real prison. It was the unexpected result of an experiment Stanford psychologist Philip Zimbardo headed in 1971. What makes this scene so disturbing is that the twenty-one prisoners and guards were really law-abiding college students, all carefully screened by a battery of personality tests revealing them to be unusually emotionally mature and stable. Coin flips decided who would be prisoner and who would be guard. The experiment then began on a quiet Sunday morning, when real police officers rounded up the prisoners. Once booked, they were jailed in makeshift cells in the basement of Stanford University's psychology building. So began the participants' lives as guards and prisoners.

Zimbardo recalls how surprisingly easy it was to elicit such disturbing behavior from a group chosen to be as normal as possible. But what Zimbardo found most surprising was the speed of their transformation. By the sixth day, guards and prisoners alike were acting so bizarrely that Zimbardo had to cancel the study, far short of the planned two weeks. Once debriefed and given the chance to gain some perspective on their behavior, the students were shocked by how far they had sunk. Reviewing their journal entries, many of the "guards" couldn't believe the hatred of their words, which translated into brutal treatment of the prisoners. Likewise, the prisoners had been so overwhelmed that when they demanded consultation with a priest, they introduced themselves using their prisoner number instead of their real names. And instead of demanding that the experiment be stopped, through sobs they begged the priest to call their parents to get a lawyer and bail them out. For everyone involved it was a traumatic experience that would never pass a university's human subjects committee today.

As unsettling as this experiment was, it has been even harder to figure out just what it reveals about us. According to one view, the prison experiment simply removed society's coercive forces, unmasking the students' inborn sadistic impulses lurking just beneath the pretense of polite civility. If so, then its lesson—really its warning—is just how fragile our social order is. The great champion of this strikingly modern view was Freud, who saw us as a bundle of instincts and filled with a warring sexuality. This image posited that we are propelled by urges beyond our control and understanding, making us ruthless, aggressive, and antisocial. Far from shaping who we are, civilization sits like the lid of a pressure cooker atop these ancient impulses, redirecting our base instincts so that we can live together without killing each other.

Others, including Zimbardo himself, would say that the experiment reveals the extraordinary power of culture and its social roles to shape who we are. Perhaps, then, the students had simply put on new, socially constructed masks and almost instantly made a different reality, just as we all juggle different social masks as worker, parent, or friend.

These competing explanations echo an age-old tension. From *Tom Jones* to *Trading Places,* we've argued whether we are the product of our history or our biology. This is, of course, the nature/nurture debate. A glance at today's headlines is all you need to convince yourself that this debate is not over.

The Modern Battle for Human Nature

More than two thousand years ago, Plato and Aristotle debated whether virtue and other human traits were inborn. Yet the rise of evolutionary theory in the nineteenth century radically altered the terms of the debate. In the early years following Darwin, many researchers culled an extreme nature view from Darwinism, revealing how volatile the intrusion of biology into human affairs can be. Darwinism was also twisted to support a crudely racist interpretation of nature over nurture. Its proponents argued not only that human traits were inborn, but also that humans belonged to different races, each with its own set of traits. Since each race supposedly represented different species at various evolutionary stages, the traits of each race could be placed along a scale of increasing worth, with the British gentleman unsurprisingly perched at the apex. In the United States, the times conspired to create a flash point in the nature/nurture debate, as the American Civil War broke out just a few years after Darwin published *The Origin of Species*. Social conservatives defended slavery with the evolutionary argument that it didn't differ from the commercial exploitation of other species. Darwin and his colleagues strongly opposed these racist views, but the plundering of evolutionary theory in the service of a racist social agenda was all but complete.[5]

Extremism inevitably invites a backlash, and Social Darwinism, with its emphasis on biologically determined traits and group differences, was no different. This backlash coalesced into cultural relativism, an environmental approach to human behavior that would dominate much of the twentieth century. Unfettered from biology, human behavior and individual differences were accounted for exclusively by social customs, habits, and socialization.[6] According to this nurture view, human nature exists only to the extent that it represents our potential to be shaped by our culture. Indeed, even to this day, the mention of biology in the social sciences often results in charges of scientism and ulterior agendas.

It wasn't until the mid-1970s that a major attempt to reintroduce biology into human affairs would be undertaken. This time, the Harvard entomologist E. O. Wilson was the most visible advocate. Most of his 1975 scholarly book, *Sociobiology,* was devoted to social insects and wolf packs, but Wilson included a final, speculative chapter on human society and biology, suggesting that human society could be understood in the same biological terms as any other social animal. Despite the brevity of his speculations, Wilson's sociobiology collided with the social movements of the 1970s. His Harvard colleagues Richard Lewontin and Stephen Jay Gould published a scathing, ideological review of *Sociobiology* in the *New York Review of Books,* cosigned by members of the Sociobiology Study Group, which was affiliated with the leftist group Science for the People. In their review, they accused Wilson of a pseudoscience that tends to "provide a genetic justification of the status quo and of existing privileges for certain groups according to class, race, or sex."[7] Continuing, they linked Wilson's thinking to "the eugenics policies which led to the establishment of gas chambers in Nazi Germany."

Thus branded a right-wing extremist, Wilson became the lightning rod for social movements, as angry mobs halted his talks and even physically attacked him during his 1979 address to the American Association for the Advancement of Science.

What had Wilson done to invoke such strong reaction?[8] In bringing together biology and evolution to understand human social behavior, Wilson viewed us through the lens of behavioral ecology, which explains animal behavior as adaptations to specific environments. Thus, Wilson argued that our behavior must be highly adapted, like that of bees going about their business or the cog-and-wheel precision of an ant colony. As you might imagine, this view had some immediate social and political fallout. For example, Wilson argued that feminism flies in the face of evolution, thereby igniting a battle with Gloria Steinem, who soon appeared on network television insisting that the scientific search for genetic differences between male and female brains be stopped. Sociobiology quickly became the forbidden fruit of the politically incorrect.

Invasion of the Body Snatchers

Wilson's sociobiology was built on what is known as the Modern Synthesis of evolutionary biology, which combined two powerful ideas: Darwin's struggle among individuals and the genetic transmission of traits. Originally, Darwin thought that evolutionary struggles might be played out among competing groups. When Darwin read the economic writings of Thomas Malthus—in particular his *Essay on the Principle of Population,* which explored the misery caused by overpopulation and the scarcity of resources it creates—he switched from thinking about competition among groups to assessing competition among individuals. This was a revolutionary move, and Darwin realized that the resulting notion of natural selection meant that individuals must differ in ways that make some more fit than others. The traits that make individuals fit must somehow be passed on, selected by evolution. Although we talk of genes and evolution in the same breath today, Darwin made his theory of evolution known at a time when the physical basis of inheritance was a matter of speculation. In fact, Darwin recognized that the Achilles' heel of his theory was the lack of a good mechanistic explanation for inheritance, which may have been a reason why he delayed publishing his theory until some twenty years after returning from the Galápagos Islands. Yet Darwin's theory wasn't entirely out of step with his times. Victorian England was enamored with the idea of progress. In an optimistic boom time, the theory of evolution suggested that life was progressing toward ever more perfect forms. Evolution's goal, then, was the perfectibility of life itself. Evolution became the links that held together the great chain of being, with humans at the summit.

The twentieth century was not kind to the idea of progress. Two world wars were enough to cast doubt on any ideas that history involved progress. So, too, biologists abandoned any notions that evolution has ultimate ends. Evolution became a blind force of nature, surely changing but not directed. It meandered as contingently as the accidents of history. Without ultimate goals or any grand design for human life, a more pragmatic directive emerged. It would be here, in the synthesis of genetics and evolution, that the starkest and most disturbing answer to who we are would emerge. As genetics and evolutionary theory fused into what is called the Modern Synthesis, evolution was reduced to a single precept: genes getting copies of themselves into the next generation. Inside all of us, then, are the genetic victors of a long and pitiless struggle.

Governed by this precept, evolution can be a nasty business. Consider the spotted hyena. Unlike other hyenas, spotted hyenas are born with strong necks and jaws, along with fully formed teeth: long canines and gripping incisors. This odd fact turns ominous with the indication that female spotted hyenas usually give birth to twins. In the dark of the den, the firstborn attacks the secondborn within minutes of birth, often before it has left the amniotic sac. The secondborn, endowed with similar armament, fights back. Whether a single sibling emerges from the den, or whether there is a stalemate, is a testament to the ruthless inborn imperative to survive. So much for brotherly love. It makes you glad you're not a hyena, until you realize that similar, but more subtle, strategies might be programmed into you.

The shift to a gene's-eye view completed the image of humans that Freud had begun some fifty years earlier.[9] Whereas Freud argued that our motives are concealed to us in the depths of our unconscious, the gene's-eye view went one step further, locating the cause of our actions in a struggle for genes to survive.

What made this new approach different was its idea of how we fit into this struggle. In 1976, the Oxford sociobiologist Richard Dawkins proclaimed that people aren't the real actors in the struggle for survival.[10] Each of us is here today and gone tomorrow, too fleeting to figure in the struggle at all. In place of selfish individuals, Dawkins gave us one of our most striking and disturbing metaphors: the selfish gene. Only genes have a real shot at immortality by passing copies of themselves on to the next generation. "Selfish" genes that produce cunning, manipulative behavior have a better chance of doing so than "nice" genes that produce dupes. Surrounded by selfish genes, benevolent genes don't stand a chance. And so inside you are immortal genes once carried by distant ancestors who were more cunning at survival and manipulation than their cohorts. Shakespeare, then, didn't get it quite right when he lamented Caesar's turning to clay to stop a hole to keep the wind away, as Caesar's genes might still be among us. Nor, according to this modern image, is our essence either that of an angel of reason or that of a poor forked animal. Instead, we are just a gene's way of making more genes. As Dawkins so starkly puts it, "DNA just is. And we dance to its music."[11]

What's more, according to Dawkins, your genes are manipulating even you. Genes that make you think *you* are calling the shots are at an advantage. Even the most sincere-looking act of charity is nothing but an unconscious attempt to ingratiate yourself to others, making it more likely that someone will want to help spread your genes. Your genes disguise your true motives because if you knew what really made you tick, you'd be so repulsed and alienated that you'd be useless to your genes. Perhaps the fitting epitaph for this twist on the modern image, then, would be Groucho Marx's quip that he wouldn't want to be in any club that would have someone like him for a member. Even more depressing, most of us don't have even this flash of insight. Thus do genes stealthily ensure their immortality. So much for know thyself.

The March of the Genes

Within the last decade, advances in genetics that even insiders couldn't have predicted have transformed a once esoteric science into an applied technology. The capstone to it all was the race to sequence the human genome between Celera Genomics, a private company, and the international Human Genome Project: the largest biological undertaking in history. With the initial sequencing and analysis of the human genome now complete, the goal of genomics is the holy grail of life sciences: to decipher the more than thirty thousand or so genes residing inside every one of us.[12] It is difficult to fully comprehend the implications of these advances. In the future, the recipe for making humans will likely be manipulated easily. If so, then genomics will decipher how genes relate to human individuality. In fact, every week it seems some new study reports a gene associated with another facet of ourselves, including shyness, depression, thrill-seeking, sexual orientation, violence, and intelligence.

In many ways, the world changed on July 5, 1996, with the birth of the now-famous Dolly, the world's first cloned mammal. Overnight, what the majority of scientists thought was distant science fiction became a reality. As captivating a symbol as Dolly was for the power of genetics, another event some two years later may prove to be even more significant. On July 22, 1998, a research group in Hawaii revealed that they had produced fifty carbon-copy mice spanning three generations. These mice clones represent an important advance because they are the result of a cloning method that's much more reliable than the one that produced Dolly (it took the Scottish group nearly three hundred attempts to create Dolly). The mice results proved that cloning technologies could rapidly improve. With clear market incentives, human cloning won't be far behind. Indeed, the Italian fertility doctor Severino Antinori has announced plans to clone humans, and a recent study suggested that human cloning might actually be easier than cloning sheep.[13] As Pierre Baldi, director of the Institute for Genomics and Bioinformatics at the University of California, Irvine, notes, from a technical point of view, as a reproductive technology, human cloning may turn out to be easier than today's in vitro methods.[14]

The implications are clear enough. Some of the deepest held beliefs about what it is to be human—convictions often rooted in tradition and religious belief—are no longer beyond the scope of scientific scrutiny. Since clashes rooted in competing self-conceptions have sparked wars, it's no surprise that recent glimpses into a genetic self-conception bring such turbulent debate. Cover stories in virtually every major newsmagazine in the last year have asked searching questions about what form our genetic selves might take.

The potential impact of genetic technologies to shape both our daily life and our self-conception shouldn't be underestimated. Just as twentieth-century physics overturned our intuitive understanding of the physical world around us, so gene research indicates that our intuitive understanding of and explanations for human behavior may be next. Indeed, the doing of science begins with accepting that closely held intuitions may have to fall by the wayside. Perhaps the earth is not at the center of the universe. Perhaps time does not unfold at the same rate everywhere. Perhaps who we think we are is just another flawed theory. Consider, for example, our notion that we choose our actions freely. With this assumption we have constructed a legal system that holds adults—whom we assume have reached the age of reason—

accountable to the law. Children and the insane are excluded because we believe they are incapable of deliberative choice.

What if findings about our genetic selves undermine this self-conception? The rise of so-called Twinkie defenses and other biological gambits are just the latest signs that the assumption of free choice underlying the law is under stress and scrutiny.[15] If the propensity to transgress society's laws is rooted in the genes some individuals inherit, or if it's just one more evolved survival strategy wired into our brains, could crime be considered a disease of the mind? Does a disease model of crime suggest that the solution to social ills lies in altering human genes—an idea that has had notorious implications in the past?

The implications of genetic research extend far beyond our legal system. The specter of a comprehensive psychological and health profile chart hanging off a newborn's hospital crib, along with a DVD containing that infant's genetic code, may lurk right around the corner. How could our interactions with that newborn not be affected by the information on that dangling chart and DVD?

As geneticists made unprecedented advances, the idea that we are just another animal for study—and that who we are is intimately linked to our genes—no longer seemed as radical as it did in the social environment that received Wilson's *Sociobiology*. Thus, evolutionary thinking began to figure again in how we conceive of ourselves. Indeed, according to a recent poll, 68 percent of Americans believe that a person's behavior is shaped both by genetic makeup and by life experiences, and they reject arguments that either factor alone determines a person's destiny.[16] During the 1980s, Wilson's sociobiology reappeared, though the advocates of this new evolutionary account of human behavior avoided association with sociobiology and instead dubbed their new framework evolutionary psychology. Much of the popular appeal of evolutionary psychology no doubt lies in its ready explanations for everything from why we manicure suburban lawns and gorge on potato chips to why some men take a game of pickup basketball with deadly seriousness. An article even appeared in the *Harvard Business Review* applying evolutionary psychology's lessons to life in the corporate tribe.[17] But its most famous and titillating musings are on differences in reproductive strategies between males and females, all of which are as likely to appear in scientific studies as in *Cosmo* articles on getting and keeping a mate.

Although evolutionary psychologists adopted much from Wilson's sociobiology, one point of departure was in abandoning sociobiology's view of human nature as an adaptation to modern life. Instead, evolutionary psychologists argued that our brains were adapted for life in the ancient past in what they call the "environment of evolutionary adaptation," which we'll just call the "ancestral environment." This was the way of life on the African savanna, where humans are believed to have spent 95 percent of their evolutionary past as hunter-gatherers in nomadic bands. Sociobiology's pessimism that the status quo reflects a genetic reality becomes more fatalistic in the evolutionary psychology model, where egregious behaviors such as rape are explained as evolutionary adaptations.[18] According to evolutionary psychology, "our modern skulls house a stone age mind,"[19] our rise from Stone Age creatures to computer age inhabitants has been too rapid for our behavior to catch up, making us genetic misfits in the modern world.

Out of Switzerland?

Evolutionary psychologists often liken the brain to a Swiss Army knife. According to this metaphor, the brain is a collection of special gadgets, each chosen by natural selection because it conferred a selective advantage on its owner out on the distant savanna. Over the eons, more and more gadgets were added, until the base model brain of our ape-like ancestors became the deluxe human model, replete with hundreds of mental corkscrews and cognitive tweezers. Whereas the Swiss have managed to put a dozen or so blades on their knives, evolutionary psychologists speculate that evolution has put literally thousands of specialized circuits into your brain, from cheater detectors to friendship modules.

Although uncovering who we are requires seeing ourselves in the light of evolution, evolutionary psychology, which advertises itself as

being based on biology, relies on indirect arguments, abstract principles, and more than a little creative storytelling. It never makes direct
contact with the brain to go beyond metaphors and examine directly
how evolution operates on the brain. And in genetics there is a growing
realization that the connection between most genes and behavior is
enormously complex and subtle, requiring us to fathom the connection
between genes and the brain that underlies our every thought, feeling,
and action.

Taking the Next Step

Understanding who you are means making a final, crucial synthesis. It
means fathoming how the forces of evolution operate on brains and
how those forces shaped the human brain. As we will examine, a major
route to evolving brains is via altering their development through
changes to developmental genes.[20] Thus, uncovering the mystery of
who we are requires understanding how you become who you are as
your brain develops over the course of a lifetime. Understanding these
themes, however, has proven to be a challenging enterprise because
the brain was not engineered by human design, but by the vicissitudes
of evolution. John Allman, our colleague at Caltech, likens the brain's
evolution to an old power plant he visited in San Diego.[21] He noticed
that several generations of systems for regulating the power were being
used simultaneously, including an array of pneumatic controls that
opened and closed tiny valves, a system based on vacuum tube technology, and several generations of computer-based systems. When he
asked why there were so many control systems, he was told that the
demand for continuous power generation prevented the plant from
ever being shut down for a complete renovation. The progressive over-
lay of technologies was integrated into one system, each depending on
the previous ones, for the seamless generation of electrical power.

So it is with the brain. Each generation must be able to survive
while nature tinkers with innovations. Some are successful and are
passed on to future generations, but others fall by the wayside. After

many generations have passed and new species have evolved, a particular component of the brain may be inextricably linked with others
that have come before and after it. This makes it much more difficult
to unravel how anything works in the brain, and until a few years ago
progress had been slow.

Even as recently as a decade ago the brain remained too shrouded
in its immense complexity to allow us to answer the most basic questions about ourselves. But neuroscientists have learned more about the
brain in the last decade than in all previous history. Pioneering new
brain-imaging techniques are giving us our first glimpses inside the
human brain at work. Researchers are also beginning to decode the
chemical soup inside the brain, uncovering the mechanisms of love
and hate, of personality and emotion. So, too, a revolution in computers has given us powerful new tools for understanding the swirling activity of thousands of brain cells that underlie our every thought and
action. To see how far we have come in such a short time, let us look
inside one brain function: our ability implicitly to learn important patterns without even being aware of what we have learned.

Follow the Spot

In a dark room, a spot of light was flashing briefly onto a screen, ap-
pearing in a new, random location twice every second. The game was
to move your eyes to each new location as quickly as possible. This
low-tech computer game was a summer research project at Cleveland's
Case Institute of Technology participated in by Terry, who was about to
become a freshman in the summer of 1964 (Case later joined Western
Reserve University to become Case Western Reserve University).
Most of the summer was spent programming an online lab computer
that controlled the lights and measured reaction times.

On average, it takes approximately two hundred milliseconds, a
fifth of a second, to begin an eye movement after the appearance of a
light, and then another fifty milliseconds to rotate the eyes to the new
position. As the measurements rolled in, something odd was happen-

ing. Subjects' reaction times to some of the lights in the sequence were getting shorter and shorter. After thirty minutes of practice, the eyes were arriving on some targets just *before* the light went on. This can only happen if the decision to move the eyes to that target had occurred well before the light had come on—that is, the brain was predicting where the next light would turn on. Was this evidence for extrasensory perception?

Although the sequence of lights looked random, it contained a repeated sequence, such as 1-5-3-1-4, interspersed among other sequences that really were random. It was during those repeated sequences that the eyes began to predict the position of the next light. The real surprise was that subjects were not aware of the regularity in the sequence. The brain can detect, and correctly respond to, regular patterns in the environment without your conscious awareness of them.[22]

Back in 1964, virtually nothing was known about how the brain learned these sequences or where in the brain they were remembered. Many of the first steps toward addressing these issues involved studying people who had lost their ability to learn sequences implicitly after having suffered a brain injury.[23] Some persons with Huntington's disease and Parkinson's disease are impaired at this task. A person suffering from Huntington's disease generates exaggerated movements, whereas someone with Parkinson's disease has trouble initiating movements. These diseases damage different parts of the basal ganglia, a mysterious part of the brain that is thought to be involved in generating sequences of movements. The basal ganglia will be discussed later in the context of an ancient learning system that is also the focus of addictive behaviors. Whereas these patients had lost this learning capacity, it is preserved in amnesiacs, who have lost the ability to recall explicit facts and events following damage to other parts of their brain. Disentangling differing patterns of breakdowns following various brain injuries remained until recently the preeminent way to identify the brain basis of human thought and behavior.

The advent of functional brain-imaging technologies signaled an important advance in brain science.[24] Using these techniques, Gregory Berns, working at the University of Pittsburgh with Jonathan Cohen and Mark Mintun, examined normal subjects in a sequential button-press task in which the sequences were generated by fixed probabilities.[25] Analogous to the results Terry saw thirty-three years earlier, the reaction times decreased even though subjects were not aware of the regularities in the sequences. When there was a subtle and unperceived change in the probabilities that generated the sequences, the reactions times increased, again analogous to earlier results. Berns and his colleagues, however, could see inside subjects' brains to watch changes in brain activity. As subjects learned the task, increased brain activity in the prefrontal cortex and the basal ganglia signaled the involvement of these structures in implicit learning. One of the activated areas of the prefrontal cortex was the anterior cingulate, which is involved in monitoring task performance and helps you focus your attention. The anterior cingulate cortex is a fascinating part of the brain that we will revisit in the next chapter. Although many parts of the brain are directly involved in this task, converging evidence points to the basal ganglia as the primary site where sequential information about the world is learned implicitly.

In addition to emerging experimental techniques like brain imaging, new theoretical approaches to understanding the function of brain systems provide new ways of understanding the brain's complex workings. Before performing his brain-imaging experiments, Berns had worked with Terry at the Salk Institute on building computer models of the neural networks in the prefrontal cortex and basal ganglia that underlie sequence learning. A surprising conclusion of these modeling studies was that learning in your basal ganglia may be similar to learning in the bee brain, suggesting that a great diversity of brains in nature may share important properties, as we will see in chapter 5.

We have learned a great deal about how the brain recognizes important regularities in the world, much of it surprising, such as the commonalities in learning across brains as different as those of the human and the bee. Discoveries regarding how the brain works often surprise us and promise to offer revealing insights into our nature. This is a consequence of a law Leslie Orgell, an exceptionally smart colleague at the Salk Institute who studies prebiotic evolution, has dubbed Orgell's second law: Nature is more clever than Leslie Orgell. Even sensible ideas can be wrong, so we need to be wary of any plausible proposal that does not have converging sources of experimental evidence behind it. The first source is observation of behavior under carefully controlled conditions; next, detailed measurements of brain activity that allow hypotheses to be formed on how the behavior is generated; finally, computational models that link these observations to provide insights and explanations that are often counterintuitive.

A new discipline known as "cognitive neuroscience" is beginning to put these pieces together. It is giving us our first glimpses into the most supreme of mysteries—how a brain makes its remarkable journey across life, over time shaping a thinking, feeling, unique person. These findings are unlike anything before in science. These are not answers to what lies in the farthest reaches of the universe or in the smallest specks of matter, as profoundly as those answers have changed our world. These new findings give the first answers to the most basic and perplexing questions all of us ask as we search for meaning in our lives and in the world around us. A hundred years ago cognitive neuroscience might have been called the science of the soul. Based on unprecedented advances of cognitive neuroscience, the outline of a view we call "cultural biology" is taking shape, forming the new science of who we are.

Making 2
Connections

ichael Dickinson, a fly researcher at the University of California, Berkeley, was wowing the five hundred physicists, mathematicians, biologists, psychologists, and engineers who were gathered at the international Neural Information Processing Systems (NIPS) Conference on November 27, 2000, in Denver.[1] Dickinson was describing robofly, a mechanical model of the fly flight system, which he had built to unravel the mystery of how flies fly. The aerodynamics of the fly's wing motion had a tricky twist that he had captured with a high-speed camera and replicated in his scaled-up mechanical model. In order to follow the flight path of a fly, he built a fly-o-rama, an arena for flies with cameras to record their flight trajectory. To understand the neural systems that control the flight trajectory he recorded electrical signals from tiny neurons in the visual and motor systems of a fly in response to movement of the background.

The meeting was unusual in the diversity of the gath-

ered scientists and engineers. Physicists typically go to physics meetings, biologists to biology meetings, each meeting focused on a narrow
topic. Each field of science has its own specialized language and background knowledge, which makes communication across disciplines
difficult, often as great as that between different cultures. What all
these tribes of science and engineering share, however, is the accep-
tance of nature as the ultimate arbiter, through a long tradition of
experiment, theory, and practice, and the common language of mathematics and computation.

The NIPS conference was started in 1987 in the midst of the neural network revolution.[2] It had become clear that the problem of understanding how brains work would require cooperation among
scientists from a broad range of disciplines. Engineers wanted to ex-
ploit the exceptional capabilities of biological sensory and motor systems, and brought to the table the ultimate proof of principle: Build it
to see if it really works. In Dickinson's case, this meant seeing if his
theory would literally fly. During the first few years at the NIPS meetings, engineers and biologists had a hard time understanding each
other's talks, but gradually over the years something remarkable happened—the engineers began to talk like biologists and the biologists
began to appreciate the engineering perspective. More important,
they began to collaborate with each other on projects that could not
have been tackled alone.[3] New journals were founded to publish their
discoveries.[4] Their students absorbed these results and did not have
as much trouble following all the talks. Dickinson is one of these new
hybrid scientists, equally at home in biology and engineering. It was
not always this way.

Thinking About the Brain

In 1979, Nobel laureate Francis Crick wrote an article for *Scientific
American* titled "Thinking About the Brain."[5] Crick, with James Watson, had set the stage for the genetic revolution in 1950 by discovering the structure of DNA. Since 1959, at the Salk Institute for

Biological Studies in La Jolla, California, he had been thinking about how brains work and had come to the conclusion that the secret to understanding the brain was in neuroanatomy, the study of how the brain is wired. There are thousands of different areas in the brain, each with a unique set of neurons and with highly specific sets of connections between them. Neuroanatomy is one of the most difficult parts of neuroscience to master, filled with terms for brain areas such as *substantia innominata* and pre-Botzinger complex,[6] made more difficult because the names of the same brain regions change in different species.[7] Moreover, compared with the elegant simplicity of the structure of DNA, the tangled wiring in the brain is a nightmare. It was a surprise to imagine that Crick could find theoretical inspiration in such complex circuitry.

The same year in which Crick published his thoughts about the brain, a meeting was held at the University of California at San Diego. The meeting brought together a small group of researchers from computer engineering, psychology, neuroscience, physics, and statistics, and would set the stage for the neural network revolution.[8] For nearly twenty years, scientists and engineers had been writing computer programs that used rules and deductive logic in their theories, seduced by the mathematical power of these formal systems. Although humans can follow rules and, with training, make logical deductions, this is not the way that neurons in the brain compute. Although these programs could solve problems within narrow expert domains, this approach had failed to produce a general artificial intelligence or achieve a deeper understanding of human behavior. In contrast, networks of neurons have powerful pattern-recognition abilities and can store information, imperfectly, as associative memories. Formal neural network theory, however, was at that time unpopular and underappreciated;[9] the most powerful network architectures and learning rules had not yet been discovered when the 1979 meeting took place, nor had the implications of neural networks for human thought been explored. But the participants at that meeting, as well as their students and colleagues, would over the course of the next decade make many of the key discoveries showing that neural networks were exceedingly powerful computational architectures, vindicating Francis Crick's intuition that the connections between neurons were a key to understanding the brain.[10]

At the time of the meeting, Terry was a postdoctoral fellow in the Department of Neurobiology at Harvard Medical School and had just finished a physics Ph.D. at Princeton University. Diving into neuroscience meant giving up the neat, elegant mathematical theories of physics for the scruffy, decidedly nonmathematical world of biology. As an example of this difference, when Leo Szilard,[11] the Hungarian physicist who invented Maxwell's demon and patented the atomic bomb, went into biology in the 1950s, he had to change his thinking habits. He was able to think through physics problems while soaking in long hot baths, but when he thought about biology problems, he had to get out of the bath every few minutes to look up a fact. Evolution has created an enormous diversity of creatures and molecular mechanisms that cannot be deduced, as much of physics could be, from first principles. This is because much of biology is as idiosyncratic as history and about as predictable as the existence of Napoleon. The shining exception is DNA, the universal alphabet of biology with which all of life is writ.

The neurons that make up the brain are made from the same molecules and have the same molecular mechanisms as the cells in the liver and heart, but neuroscience also depends on insights from computing, since the brain is in the business of gathering, evaluating, storing, and acting on information about the external world as well as the internal organs.[12] However, the signals used by the brain to process information are quite different from those found in digital computers. For example, although neurons use electrical signals to communicate from one part of a neuron to another via wirelike structures called axons (axons can be quite long—up to a meter in length), they use chemical signals to communicate with each other at junctions called "synapses." Why is this? In part, the reason is that cellular machinery was inherited from ancient signaling systems used by primitive cellular organisms like algae and fungi to send chemical "smoke signals" to each other. Hundreds of genes are involved in this process, including proteins that synthesize the chemical messengers, other proteins that concentrate the chemicals into packages called vesicles and release them from the sending cells, and receptor proteins on the receiving cells that selectively bind to the chemicals and convert the signal back to an electrical form. Nature tends to reuse old machinery for new purposes and, as a consequence, our brain looks like a Rube Goldberg machine.

Terry was trained as a theoretical physicist, so it came as a revelation when he peered through a microscope at a living neuron for the first time, at a summer course in neurobiology at Woods Hole, Massachusetts, and wondered at the beauty of what he saw. The experience was a little like looking through a telescope at the vast universe, but now the infinitude was concentrated into a tiny package that you could poke with a pair of tweezers. Later, as a postdoctoral fellow in neurobiology at Harvard Medical School, Terry worked with Steve Kuffler, a founder of modern neurobiology, on the transmission of signals at synapses in sympathetic ganglia of the bullfrog. This was also a revelation, since it taught him that time, as well as space, was used by neurons in marvelously complex ways. The gap between neurons at a synapse is so small that it takes only a fraction of a millisecond for the chemical transmitter released from the sending neuron to be detected at the receiving neuron. Most synapses in the brain work on this fast timescale, and acetylcholine acted as a similarly fast chemical messenger at the bullfrog's sympathetic ganglion synapse. However, the sending cell also released a different type of neurotransmitter molecule, luteinizing hormone-releasing hormone (LHRH), a peptide that had a much slower effect on the receiving cell. The influence of LHRH reached its peak one minute after stimulating the nerve and lasted over ten minutes—ten thousand times slower than the action of acetylcholine. This was a valuable lesson: Unlike a digital computer that runs at one clock speed, each part of the brain runs at whatever speed is appropriate for the problem at hand. For the sympathetic ganglia, this included both the millisecond timescale and the slower timescales of other bodily organs, which can take minutes to alter their functions.

Who would have guessed that so many lessons could be learned from a ganglion cell in a bullfrog? The beauty of neurobiology is that even the humblest neurons have almost all of the machinery found in the fanciest ones, and you can find the answer to a question more easily if you can study it in a simple, approachable system rather than in the most complex ones.[13]

Terry moved to Johns Hopkins University in 1982 to set up the first computational neuroscience laboratory in the world in the Thomas C. Jenkins Department of Biophysics. This was an exciting time in neuroscience, since techniques from molecular genetics were revealing the machinery that neurons use to signal each other and change their properties, and new brain-imaging techniques were allowing brain activity to be recorded in humans.[14] It was also during this period that the computational power of neural networks was demonstrated, and digital computers become fast enough to simulate realistic models of the brain itself.[15] The stage was set to tackle the most challenging problem of all—who we are—using all of the tools of molecular, systems, and computational neuroscience.

Our Journey Begins

As active researchers in cognitive neuroscience, in this book we serve as backstage guides on the trek to discover who we are. The journey mirrors one we began together a little over a decade ago, when Steve joined Terry's lab, just as it relocated from Johns Hopkins University to the Salk Institute for Biological Studies, one of the world's premier research centers in neuroscience and molecular biology.

Steve came to the laboratory interested in brain development. The work being pioneered in Terry's laboratory was well suited for exploring how billions of nerve cells somehow spin a mind. During those early years, Steve learned how to appreciate the multiple levels of brain organization, from studying individual cells in live slices of rat brain with confocal microscopy and the tools of neurophysiology to conducting computer simulations. Today, our work involves using brain-imaging techniques on human subjects, computer simulations, neurophysiology, robotics, and traditional "wet" neurobiology.

Key to this new approach was the sense that traditional disciplinary boundaries were an obstacle to understanding how the brain underlies our mental life. Steve's thesis committee, composed of computational neurobiologists, developmental neurobiologists, linguists, and philosophers of mind, reflected the growing interdisciplinary study of mind. Rather than divide and conquer, understanding the mind required uniting varying views and disciplines. This approach was nowhere more striking than in how it shifted our understanding of development. Virtually all of the classic controversies swirling around who we are, such as the nature/nurture debate, are really questions about how your brain and body develop. And this explosion of knowledge was literally reshaping the entire developmental landscape. How could it not have profound consequences for our quest to understand who we are?

Contrary to the seminal idea of evolutionary psychology that our brain is a bundle of instincts dictated by our genes, the new understanding shows something entirely different. We increasingly realized how the world we are immersed in literally helps to shape our brain. In partnership with precisely timed developmental programs, the world helped construct your mind's circuits when you were growing up, and it continually reshapes them as you experience new things and call on new skills. Moreover, this process doesn't end at adulthood. The world stirs the chemical soup inside your head throughout your life. While we were writing this book, Fred Gage, a colleague at the Salk Institute, showed that contrary to textbook doctrine, new neurons are born in the adult brains of mice and humans.[16] What's more, the survival of these neurons in mice and how quickly the mice learn new behaviors depend on the richness of the environment—even a running wheel helps make mice smarter.[17] In later chapters, we'll explore a new way of thinking about how the world and your genes interact, one whose lesson is that you are flexible *because* of your genes, not in spite of them.

These new findings take us far beyond the theories of genetic blueprints and precise mental designs that have predominated in recent decades as well as the contrary view that humans are entirely creatures of their own cultures. They form a new image of the human core, re-placing the opposing ideas of a rigid, fixed human nature versus com-plete malleability with a much more dynamic and intriguing one. What determines both our potential and our limitations as a species? As we pondered this question, we realized that the answer lies in a power that the human brain appears to possess, something at the very heart of what makes us human.

The Riddle of Self-Awareness

Ever since Darwin knocked us off our pedestal, researchers have con-tinually narrowed the gap between ourselves and our closest relative, the chimpanzee. But our evolutionary paths seem to have taken us to different lands. After all, it was Keats, not a chimp, who once started a poem with the line, "When I have fears that I may cease to be." Chim-panzees appear to have the first inklings of self; they recognize them-selves in mirrors, for example, and many of the capacities once reserved uniquely for humans have their roots our nonhuman rela-tives.[18] But as far as we can tell, they don't ponder their future, as did Keats (who was right to have such fears, as he died a little shy of his twenty-sixth birthday). Somewhere back in time, our ancestors ap-peared to have crossed to the far shore of some mental Rubicon.[19] Our ancestors' struggles and explorations took them to a land imbued with meaning, ritual, and significance. In order to understand who we are, we have to go back to when our ancestors literally invented them-selves. There, we will seek out what happened to them, and why they crossed this mental Rubicon.

As we will discover, understanding our past is hampered by large gaps in the physical record. However, an important clue lies within us all: the structure of our brain. Somehow, our brain must differ from that of our closest relative, the chimpanzee, and these differences must underlie distinguishing capacities. This possibility struck Steve vividly during a visit to the San Diego Zoo. While observing the bono-bos, his then two-year-old son Elliot made friends with a young bonobo. A sibling species of chimpanzees, bonobos are very different from their genetic cousins as we explore in chapter 3.[20] For some twenty minutes, the two youngsters squealed at each other, played games of peekaboo, and entertained each other in the typical ways kids do. For everyone on the paying side of the glass, it was an amazing event, these two children seeming so similar and yet so different. But what in the tangle of nerve cells inside their heads will make Elliot and his bonobo friend take such different paths as they grow? Why will Elliot learn a language, develop a rich sense of self, and construct a morality based on understanding that others are people with their own inner worlds of experiences and feelings,[21] while his bonobo friend will not develop these full-blown capacities? Ultimately, the answer lies buried somewhere deep inside their brains.

In beginning to answer the central riddles of the human mystery, brain science is focusing on a region of the brain known as the "prefrontal cortex," the part behind our high forehead. Damage to regions of the prefrontal cortex affects behavior in profound ways, from altering personality to impairing our ability to plan for the future.[22] Until recently, the prefrontal cortex was mostly a vast uncharted territory. During the last decade, however, the tools of cognitive neuroscience have begun to uncover its secrets.

One of the most intriguing parts of the prefrontal cortex is called the "cingulate," and in particular the anterior part, located near the centerline in the front of the brain. Although this area measures only a fraction of an inch across, damage to the anterior cingulate cortex on both sides of the brain produces akinetic mutism: Patients remain silent and sit quietly without initiating action. Patients who recover from this condition report that they did not move because they did not feel the need to do so and that their minds were "empty" of thoughts. One day in 1991, Francis Crick arrived for tea at Terry's lab and announced that the anterior cingulate in humans may be the site of our "free will."[23] In clinically depressed patients, metabolic activity in the anterior cingulate is less than that in normal humans.[24] We will see later that the anterior cingulate is exceptionally active in tasks that require concentrated effort and may be responsible for monitoring performance. The anterior cingulate in humans also has features that distinguish it from the same region in all other mammals.

Other regions of the prefrontal cortex are involved in working memory and also activate when we are faced with a challenging task that requires planning. Based on these observations, the prefrontal cortex has been likened to an executive system that guides, coordinates, and updates behavior in a flexible fashion, particularly in novel or complex tasks.[25]

The mental capacities we have mentioned so far reflect the traditional view of the prefrontal cortex as a brain region underlying complex, rational thought. In recent years, however, it has become increasingly clear that the prefrontal cortex is also intimately connected to our emotional life and that the feelings that accompany our experiences and thoughts play an important role in both thought and behavior.[26]

The intimate connection between who we are and the prefrontal cortex has spurred brain scientists to compare the human and nonhuman prefrontal cortex. The traditional view that stemmed from these comparisons was that the human prefrontal cortex expanded relative to other brain structures, making us front-heavy. Recently, this claim has been reevaluated, and it now appears that only some portions of the prefrontal cortex have expanded in the human brain.[27] In particular, a region at the front of the prefrontal cortex known as "area 10" appears to have undergone a considerable increase in size during human evolution along with an increase in its connections to other brain regions.[28] Also, brain cells in area 10 appear to be the most complex in the cortex.[29] Although the function of area 10 remains mysterious, as we examine in more detail in chapter 6, damage to area 10 impairs an awareness of one's self. Perhaps, then, the expansion and reorganization of area 10 during human evolution led to our capacity to generate a sense of self, a possibility we will pursue throughout this book.

Developing a sense of self is a protracted process, requiring long exposure to the social world.[30] Yet it is absolutely essential to human life, for we require a sense of self to navigate the complex social worlds we inhabit. Doing so endows us with the capacity for flexible behavior, as we update our behavior to accommodate rapid social changes. We were intrigued by the fact that the prefrontal cortex is the last part of our brain to mature during development, not reaching its full function until after puberty.[31] Perhaps we literally build our sense of self as our human culture helps to build our prefrontal cortex. If this were so, then our mind would be supremely flexible not because it has somehow unfettered itself from our biology, but *because* of our biology. As counterintuitive as it may seem, the more we learn about the brain, the more we realize that its rich interaction with the world is the key to understanding our complex mental life and intelligence. The idea that nature and nurture are two competing forces is so deeply entrenched in nearly every facet of how we understand ourselves that it has obscured any possibility of a rich collaboration between these two elements. Yet our research has found that humans are the result of the most complex collaborative project in history, whose two equal partners are our biology and the human culture we are immersed in. As you will see, they form a tangled web of forces whose interconnections we are only now starting to appreciate using the dynamic tool of computer simulation.

Appreciating the far-reaching interplay between biology and culture has prompted us to call our view "cultural biology." We'll show you why cultural biology points to a very different conception of who we are than does the outdated "modern" image, and why the modern image was mistaken in some of its basic assumptions. We'll also explain why we think this new understanding offers a far more fascinating view of who we are, one containing both exciting new possibilities and lurking vulnerabilities that we should all be aware of as we shape the next millennium. To that end, our journey will go far beyond the laboratory and into everyday life to extract the answers to the most basic questions that arise as we all try to make sense of our world and our place in it.

How to Make a Human

3

The 1.6 Percent

Solution

You've just received a phone call from a friend who wants you to join an engineering start-up. Intrigued, you ask for some details. He tells you that the company is going to build computers that work like the human brain and are just as powerful. They'll be in demand everywhere, from industrial applications to the entertainment industry, perhaps even as substitute children,[1] he enthuses.

Before you tell your boss you quit, there are a few more details we think you should consider. Your brain contains more than one hundred billion treelike nerve cells called neurons. Most neurons communicate through connections known as synapses, which number around 1,000,000,000,000,000 in your brain. This nexus of connections is sometimes compared to the circuits inside a computer. That is, electricity runs down a neuron's main trunk, its axon, which branches into thousands of smaller roots. When the electricity arrives at the end of the line, it sparks the release of a chemical called a neurotransmitter. Once the neurotransmitter crosses the synapse, it clings to special docking stations, or receptors. This causes a rush of electrical activity in the cell's receiving lines, branchlike structures called dendrites, among the most beautiful of nature's architecture. In the time it has taken you to read this description of the neural events inside your head, there have been more synaptic transmission events inside your head than all the stars in the night sky.

The number of chemicals your brain uses to communicate isn't fully known. Early on, researchers suspected that there might be perhaps a handful of such chemicals and that a particular brain cell used only one kind. Today, more than a hundred have been identified. Some, like serotonin, are involved in regulating your mood. Others, like glutamate, are involved in thinking and learning. If you've ever had a headache after eating Chinese food, the suspected culprit is the additive monosodium glutamate (MSG), which might alter brain levels of glutamate. These chemicals don't work in isolation. Instead, they act in concert with one another like musicians rendering a complex score. Your brain can combine these chemicals in different ways to create a bewildering diversity of tunes, from neural Bach to biological Metallica.

Your friend might have tried to convince you that these nerve cells aren't really all that fancy. He may have told you that they are more like old-fashioned vacuum tubes than modern computer circuits, switching on and off a million times more slowly than the relays inside your personal computer. He's right that nerve cells are slow in comparison to computer circuits. But when you congregate these sluggish nerve cells in a human brain, somehow you get a computer that's over a thousand times more powerful than today's fastest supercomputer, and this is not even counting the potentially great computing power within dendrites themselves. Considering all that computing horsepower, what's really amazing is that your head doesn't simply explode in a shower of sparks the minute your brain turns on. It doesn't because your brain consumes only as much energy as a ten-watt lightbulb and is water-cooled even more efficiently than your car engine. You don't even work up a sweat thinking.

To give you a better idea of what a piece of engineering your brain is, let's benchmark it against modern supercomputers. You may have noticed that your personal computer tends to heat up a room pretty well. That's because computers pack a lot of circuitry into a small space. A real stumbling block in making supercomputers is figuring out how to cool their circuits so they don't melt. Cray supercomputers, once icons of modern industrial design, were only sleek-looking because the tons of cooling equipment they require were hidden in the basement. In fact, most of Cray's patents are for cooling systems, not computing parts. To make a computer that rivals the brain in all its features, somehow you'd have to pack a thousand supercomputers into a three-pound package that requires only about as much power as the on/off lights on a typical computer.[2] Now make it capable of withstanding not just being dropped but also the occasional blindside quarterback sack or first-round barrage from Mike Tyson. Your brain is the ultimate portable computer.

Your friend had better not put a second mortgage on his house to raise funds for his new venture. In fact, millions of dollars have been devoted to building computers that mimic facets of the brain, yet machines are just beginning to do the things we do effortlessly when we open our eyes or recognize a voice. When you figure that the brain is the result of the longest R&D project in history—evolution—it isn't too surprising that we're only scratching the surface of its power with our own relatively primitive technologies.

Before we ponder your brain's incredible power further, we need to identify a few landmarks. So far we've been talking about "the brain." But that's a bit like talking about your recent trip to Europe. Europe is a lot of things: some new, some ancient. Some of it is deservedly on the highlight list, and some of it—deservedly or not—gets skimmed over without a second thought. So, too, with your brain. The Paris and Vienna of your brain are the left and right cerebral hemispheres, typically the highlight of any tour inside the human brain. Their surface is the highly convoluted cerebral cortex, actually a two-millimeter-thick layer of brain cells folded over and over again to fit inside your head like a scrunched-up wad of paper. Unfolded, the cortex would cover about four pages of writing paper. It might surprise you to know that about half your cortex is devoted to seeing, the sense we rely on above all others. Much of the rest of your cortex is devoted to other senses and to moving your body.

We have already mentioned the important place of the prefrontal cortex in human mental life, making it the Louvre of the brain, typically getting top billing as the highlight.

Of course, what makes Paris such a wonderful city is how all the parts fit together, and the same is true of the brain. Indeed, a more apt use of the Parisian brain metaphor might be to think of the prefrontal cortex as the Pompidou Center, a piece of modern architecture in the heart of an old city. As we shall see, at the heart of who you are is a complex blend of new and old regions, a Picasso-like prefrontal cortex grounded in the old masters of more ancient brain structures, some of them so old that humans share them with insects.

In the brain, all roads to the cerebral cortex lead first to a structure called the thalamus, a traffic circle buried deep inside the brain on top of the brain stem. Your brain stem receives information from your body via the spinal cord and cranial nerves and helps maintain many of your bodily needs. It also sends information up to the thalamus, which also receives information from all the senses. The thalamus, in turn, relays all this information to the cortex.[3] These structures, buried deep inside your brain, play a key role in making you who you are. They aren't the prime tourist destinations; they are like the vitally important industrial and agricultural areas on the outskirts of Paris. In fact, one small region called the hypothalamus is arguably the most important four grams of brain you own.

Now that we have our bearings, let's return to your friend's offer, which we hope you've reconsidered. How would you ever assemble such a complicated mechanism? You might suspect that it's an incredibly carefully orchestrated process. Even computer chips, which we've seen are weak by comparison, require space-age fabrication rooms more sterile than any hospital, replete with technicians in space suits.

What's even more surprising about building brains is that there isn't a lot of raw material to work with. Or at least not a lot of uniquely human material, as researchers have learned from studies of genetic differences between humans and chimpanzees. In chapter 2, we mentioned the striking friendship that Steve's son Elliot made with a young bonobo ape at the San Diego Zoo. These two youngsters, who seem both so similar and yet so different, may have been separated by an inch-thick plate of glass at the zoo, but a smaller margin—a mere 1.6 percent—separates them genetically. Elliot and his bonobo friend share 98.4 percent of their genes.[4] There's more genetic difference between the gorilla and the bonobo than between the bonobo and us. So, all that is uniquely human rests in the slimmest of margins, a mere handful of genes.[5] Before we get too chummy with our bonobo cousins, we'd also do well to remember that while we've spread across the globe to dominate life everywhere, bonobos and chimpanzees cling precipitously to a rapidly diminishing range. Bonobos and chimpanzees appear to have a rudimentary sense of who they are—we noted earlier that they recognize themselves in mirrors, for example. But they may not have the same awareness of themselves as we do, though we should caution that there is a communication barrier between them and us that limits what we can infer about their internal world of experiences.

One possible explanation for the behavioral differences between us and our ape cousins lies in their upbringing. They might act like apes because they've been culturally deprived. Not wanting to leave any rock unturned, scientists tested this idea back in the 1930s. Winthrop Kellogg, a psychology professor at Indiana University, and his wife, Luella, raised a female chimpanzee named Gua alongside their baby boy, Donald. In the Kellogg household, Donald and Gua were treated like twin brother and sister. Gua wore clothes and shoes, was potty trained, ate the same foods as Donald, and lived according to the same daily schedule. To the Kelloggs' surprise, Gua seemed to beat young Donald on most measures. She listened to commands better, was more cooperative, even more affectionate. And something strange was happening to Donald. He seemed content to follow Gua's lead and follow after her, learning from her. By the time most children begin speaking, Gua hadn't learned a word of English. Donald, on the other hand, was fast learning chimpanzee. In particular, when he was hungry he would bark the chimpanzee food call. By around eighteen months, the effects on Donald were severe enough for the Kelloggs to terminate the experiment.

The Kelloggs had started out trying to see if a chimpanzee could be made into a human and instead had discovered how easily a human could be made into a chimpanzee. "Monkey see, monkey do" turned out to be more appropriate as "human see, human do." In fact, the next time you're at the zoo, watch the people around the chimpanzee display. Chances are, a lot more people are imitating the apes than vice versa.

Donald and Gua reveal that the difference between humans and chimpanzees has to be reflected in the genes somehow. But as we've noted, we humans don't have many new genes that belong just to us. It's a small difference, but one that obviously makes a large difference. The question is, how? How do you build a human with a handful of novel parts?

One response is to suppose that there's a genetic blueprint for the human brain. According to this scenario, the brain is a collection of specialized circuits, each chosen by natural selection, each built according to a genetic blueprint. This idea is popular with evolutionary psychologists and sociobiologists alike. According to E. O. Wilson: "the newborn infant is now seen to be wired with awesome precision" and "the connections they [nerve cells] make are precisely programmed and guided to their destinations by chemical cues."[6] The linguist Steven Pinker makes an even stronger claim:

> . . . grammar genes would be stretches of DNA that code for proteins, or trigger the transcription of proteins, in certain times and places in the brain, that guide, attract, or glue neurons into networks that, in combination with the synaptic tuning that takes place during learning, are necessary to compute the solution to some grammatical problem.[7]

Do we roll off the human assembly line equipped with standard "genesoft" operating procedures built into our mind's circuits?

These ideas are reminiscent of a landmark theory of brain development advanced by a heroic figure of twentieth-century brain science, Caltech's Roger Sperry, who received a Nobel Prize for his work on split-brain patients. Sperry thought that brain cells were inexorably drawn toward one another by unique lock and key chemical codes, like a postal system for the brain, with each nerve cell having its own molecular address.

The Post-Modular World

Although Roger Sperry's theory of development was a landmark, built on creative and technically demanding experiments on the connections from the frog's eye to the frog's brain, by the early 1970s, experimenters were chipping away at it as a theory of how the cortex develops.[8] From the late 1980s onward, brain researchers found more and more evidence that cortical development was far from a "precisely programmed" genetic plan. The new results, though, were much more intriguing than the simplistic alternative of a blank slate.

Perhaps the most striking of the new findings sprang from a classic question about the cortex. Is a cortical area, such as the rear part of the brain that processes visual information, capable of taking on a new function, such as processing tactile or auditory information, or is its destiny as a visual part of the brain sealed early on? This is a fundamental issue, because if a cortical area can take on different functions, something other than a built-in plan must be instructing it what to become. By the late 1980s, brain scientists had devised delicate procedures to transplant pieces of embryonic cortex to new regions, opening the door to answering this question definitively. Dennis O'Leary, then at Washington University and now at the Salk Institute, and his student Brad Schlagger performed the transplant experiments in rats. As the transplanted piece of brain developed, it took on the appearance of its new neighborhood. In fact, the border between transplant and host

brain was seamless. Some cortical regions appeared largely inter-
changeable, in contrast to the predictions of the precisely programmed
view.[9]

If cortical regions can take on different functions, what about brain
cells? Do neurons have genetic instructions for where they should
grow and what neurons they should contact to form the mind's circuits, thus differentiating brains among spieces? Fred Gage and his
colleagues transplanted fetal pig cells into adult rat brains. The pig
cells acted like rat cells, growing in ways appropriate to rat neurons. In
other words, for brain cells, "when in Rome" means that when in a rat's
brain, do as the rat brains cells do.[10]

That's all very well for rats and pigs, you might say, but what about
humans? For decades, brain researchers have been trying to devise
tools to probe the human brain at work without disturbing it. In the
last few years, thanks to functional magnetic resonance imaging
(fMRI), a modified form of the diagnostic tool in medicine, researchers can scan the brain's activity while it's engaged in various
mental challenges. With this tool, we're getting our first glimpse inside
the human brain at work.

Using fMRI, researchers are now probing human cases that resemble the transplant experiments. Although researchers have not transplanted the human brain, nature provides her own, sometimes tragic,
experiments that afford important insights into how the brain works.

There is a widespread belief that people who lack one sensory function, such as sight, compensate through increased capacities in other
modalities, a process known as "sensory substitution." There has been
a long fascination with questions regarding the brain's response to sensory loss, from William Molyneux's classic question in 1698 to his
friend, the philosopher John Locke, concerning whether a congenitally
blind person who suddenly recovered vision would be able to recognize
objects visually, to cognitive neuroscientists who are probing similar
questions today.

In 1996, Norihiro Sadato and his colleagues at the National Institute of Neurological Disorders and Stroke placed blind subjects inside
a functional MRI scanner. With the scanner generating images of the

subjects' brain activity, Sadato and his colleagues asked subjects to read words in Braille. As the subjects' fingers moved along the Braille dots, their occipital cortex, located in the back of the head, lit up with activity (you are using your occipital cortex, better known as the visual cortex, to read this book right now). In the blind subjects, however, the visual cortex had been transformed into one for processing touch information for Braille. The visual part of the brain could be taught to learn to read with fingers.[11]

These are fascinating results. The corkscrew on a Swiss Army knife will never transform itself into a can opener, no matter how many cans you try it on. Yet here are brains that have undergone large structural modifications, altering the organization of one area to handle information from an entirely different sense. How can this happen? Does the visual brain have a backup blueprint that gets pulled off the shelf in the event of a last-minute glitch? In other words, maybe the visual brain is really a vision and touch area, with two sets of building codes. Has evolution really built in backup plans in case one module fails? On the face of it, this seems highly unlikely, considering our small uniquely human genetic repertoire. As you will see, these and other results during the last decade reveal that, contrary to what E. O. Wilson and evolutionary psychologists suggest, cortical development is not a precisely programmed, stately tour along a prescribed path. These experiments suggest a new metaphor: Cortical development now looks like pioneer days in the American West. Incoming axons, like the settlers, follow rough trails into a world of possibility. While a complex array of molecular machinery helps lay down the trails, the activity of incoming axons after they get to their destination helps determine where they ultimately settle and put down roots.[12] Of course, there's the possibility of claim jumpers. If an underused area looks promising, why not try it out? This looks like what happens to the occipital cortex of the blind. Axons relaying touch information overrun the occipital cortex in their search for more neural turf.

As intriguing as these results are, they lead to a new mystery: If a "precisely programmed" plan doesn't account for the wiring in the cortex, what does?

Brain Informants

One possibility is that the information out in the world helps build your brain. To make this concrete, imagine the trip that light makes as it streams into the atmosphere from millions of miles away, bouncing off the objects in front of you, and then irradiating your eye. If it weren't for the fact that the world has enduring structure, and the earth is not enveloped in a perpetual fog as on Venus, you'd see nothing but the snowy image akin to the screen of an untuned TV. Objects have edges and surfaces that reflect the light in reliable ways. This is information and it's there in the stream of light hitting your eyes. The same is true of the book you're reading right now. A printer has filled the page with patterns of ink that reflect light differently from the blank page or one with random markings. When you turn on your bedside light and look at the page, a lens in your eye images this pattern of light on its back surface.[13] This information is sent to your brain, which decodes the signals into letters and words. The same is true of the sounds we emit when we fill the air with vibrating energy that your ears detect and your brain decodes as words.

Could it be that the flow of information helps build the brain? Understanding how our brains and the world interact is at the core of cultural biology. The last decade of brain research has converged on the realization that worldly information and the brain are two sides of a very complex interacting system. As we'll see, your experiences with the world alter your brain's structure, chemistry, and genetic expression, often profoundly, throughout your life.

If you spent an hour or two a day for the next three weeks learning to play the piano, you would be altering the structure of your brain. Specifically, you would alter a part of your brain that represents your body and helps move its parts in precise fashion. Envision it as a map in which neighbors are body parts, not states. When you learn to play the piano, you stimulate the region of the brain's map that represents your fingers, challenging your brain to learn the new motor skills. It responds to your budding virtuosity by expanding the brain region that represents your fingers, possibly initiating a border war with its neighbors, forming coalitions with like-minded body parts. Your fresh demands spur reorganization within your brain that can have far-reaching consequences.[14] Inactivity, on the other hand, can lead to a decrease in the representations underlying a skill, indicating that the brain is constantly adapting to external events that affect how you use it.

Even more astonishing, Fred Gage and his colleagues at the Salk Institute for Biological Studies recently showed that stimulating environments spur the maturation of new brain cells in a part of the brain called the hippocampus. This is intriguing, since the hippocampus is used for forming long-term memories for facts and events and its destruction is implicated in the memory loss that accompanies Alzheimer's disease. The idea that the brain doesn't add new cells after birth was the last of the dogmatic myths to fall to the emerging view of the dynamic, labile brain. We now believe that new challenges throughout life can spur new brain cell growth as the brain responds to the demands you place on it. In fact, using your brain, especially in the early years when you are growing, is important for retaining mental abilities in old age and reduces the risk of large mental declines in later life, as we explore in chapter 9.[15]

Brain researchers call our brain's far-reaching capacity to alter its structure and function "plasticity." Understanding its implications is beginning to have concrete benefits, from designing programs for early education remediation to creating lessons for maintaining mental vitality during the golden years. For example, brain researchers Paula Tallal and Mike Merzenich are starting to tap into brain plasticity to design computer programs that help children with reading difficulties. About 10 percent of all children have reading problems that are rooted in an inability to break down speech into small chunks.[16] Merzenich, a pioneer in brain plasticity studies, and Tallal, a leading authority on the brain and reading, combined their expertise to help design a computer program that appears to rebuild the region of children's brains that is used to segment speech. The program works by first stretching out the sounds, making it easier to hear these bits of speech, and then gradually speeding them up as the child gets more proficient.

In the not-so-distant future, similar sorts of programs might be available to take advantage of the brain's plasticity in learning foreign languages. One of the problems adults face in learning a new language is that they cannot hear some sounds of the foreign language. Japanese speakers, for example, can't distinguish between the sounds made by *R* and *L*. At birth, our brain is capable of distinguishing the sound units, or phonemes, of all human languages. As we bathe in the language of our caregivers during the first six months of life, our brains tune themselves to recognize the phonemes that are present in our native language, discarding those that aren't. Think of it as another neural turf war in which neurons activated by native phonemes get to expand their empire.[17] At the cost of not being able to distinguish the sounds of a language you'll probably never hear, your brain makes the job of understanding the language you are exposed to much easier. The only reason you can carry on a conversation at a crowded cocktail party is because, unknown to you, your brain is incredibly busy pulling the right signal out of the buzz all around you. This early tuning helps make you a competent cocktail party conversationalist.

Although it is accepted doctrine that once this tuning process happens there's no going back, there are some hints that your brain is capable of learning new sounds. But it needs to hear these sounds exaggerated in order to recognize them. Have you ever wondered why parents talk in such cutesy ways to their babies? This behavior occurs in cultures worldwide and babies love it.[18] Stretching out the sounds for babies may make their job of learning easier. One day, computer programs might do this for our mature brains, helping to rebuild them to hear the phonemes of foreign languages, so that learning a new language becomes much easier and lets you converse more like a native speaker.[19]

These intriguing findings support the emerging view of the responsive brain and are especially good news for everyone sporting older models. An older brain needs new challenges—old brains can learn new tricks within limits that are still not fully understood, as we'll examine in chapter 9.[20] This new interactive brain-and-world view leads to a most surprising lesson, but first we need to look at the role genes play.

The Fall of the Genetic Despot

Inside each cell in your body (except for red blood cells and a few other exceptions) is a command center, a nucleus. Inside the nucleus are strands of tightly wound molecules called DNA. These make up the thirty-one thousand or so genes that make you human. If you unsnarled and linked all the DNA in your body, it would stretch to the sun and back—between ten and a hundred times. In simplest terms, a gene is essentially a stretch of DNA that is a recipe for making a protein.[21] Proteins are remarkable molecules whose complex three-dimensional shapes endow them with specific functions that animate life itself. Some, for example, are the neurotransmitters that brain cells use to communicate with one another. Others are the receptors that receive these messages. Still others are the enzymes that make and degrade these messengers and keep your cells ticking.

Here's the twist. Almost all your DNA—more than 97 percent of it—doesn't contain recipes for making proteins. It seems to be sprinkled everywhere, even inside some genes, cutting them into isolated pieces. Much of this noncoding DNA seems to do nothing at all, and so has been named "junk DNA." But other stretches turn out to be really important. They're the instruction manual for how and when to activate genes, referred to as regulatory elements. When a chemical messenger protein known as a transcription factor docks to one of these regulatory elements, it changes a neighboring gene's activity. It can turn the gene on to start the protein-manufacturing process, speed it up, slow it down, or turn it off. This is termed *gene regulation*. Remember, the cells in your body all have the same genes. What makes one a kidney cell or a brain cell is the genes that that cell expresses—the ones that are turned on. Your brain expresses the most genes of all your organs, requiring an enormously complicated pattern of gene regulation.

What drives all the gene regulation inside your head? Genes are often thought of as operating independently of environmental signals, something like a despot at the helm of an empire. However, signals from the world routinely tell genes what to do. To see what dramatic forms this can take, consider the female prairie vole, a mouse-sized rodent found in the American Midwest. She does not reach puberty at a predetermined age. Instead, she reaches puberty only after she has been exposed to a chemical signal from the urine of an unrelated male. While that might sound bizarre, smell information plays as key a role in rodents as visual information does in us. A whiff of this signal sets off a cascade of gene expression in the female prairie vole, transforming her into a sexually receptive mate within twenty-four hours—not exactly a dozen long-stemmed roses, but it works for prairie voles. The mystery behind this mechanism is solved by the knowledge that the prairie vole is a monogamous species. The female forms an enduring bond with the first male she falls for and will only have eyes for him for life. Holding off her sexual maturity until she encounters an unrelated male helps ensure that she doesn't form this bond with a relative, which would lead to inbreeding.

Hormones like estrogen and testosterone can bind to receptors on the cell surface, pass straight into a cell's nucleus, and change gene expression. But they aren't the only way to get into a cell's genes. We now know that learning and other experiences with the world routinely alter human genetic activity. Even more intriguing, experiences such as learning turn on a specific kind of gene. Some genes make transcription factors, which as mentioned are proteins that regulate other genes. Understanding these powerful genes, then, is of paramount concern, because some of them, called oncogenes, have gone out of regulatory control in cancer.

Learning and other experiences alter the activity of these genes. This interplay of environment with genes routinely triggers a cascade of gene activity, often leading to enduring cellular changes that underlie our brains' construction and plasticity throughout life. Here are some examples: Environmental stimuli turn on genes that make some of the receptors your brain uses to communicate. Visual experience turns on genes for the development of the visual part of the brain, turning off genes that keep it immature. Environmental stimuli also turn on genes that govern the branching and growth of brain cells. Parental behavior such as touch turns on genes that help offspring handle stressful events.

Genes, then, are the tools experience uses to change the brain's response to new demands in new environments. You are flexible because of your genes, not in spite of them. Understanding more about how genes contribute to making you who you are doesn't tilt the scale toward genetic determinism, as has been commonly assumed. Instead, it requires that we see the brain and world as an interacting complex system, a web of forces.

Don't differences in our genetic makeup pose limits on us? Indeed, there are limits to how far we can respond. But is there any physical device that doesn't have limits? Your car can go only so fast, but that doesn't mean it's not flexible. Steering wheels, accelerators, and brakes make cars responsive to varying conditions. Think of the genes in your brain in the same way. We may be driving slightly different models, but every model comes equipped with an accelerator, a steering wheel, and brakes.

Far from a being a dictatorship, our genes are more like an inner council with a large constituency, open to the needs of other cells and even to foreign influences. Ironically, just as authoritarian governments were falling everywhere around the world in the late twentieth century, so too were they falling in the brain sciences. The last wall, the one keeping the world from genes, quietly fell, too. The brain is a democracy and the world gets a vote.

The Key to Making a Human: The New View of Development

With this new model of mind in view, let's get back to the question of how to build a human. We've noted that there's not much genetic elbow room between you and a chimpanzee, yet the differences between a chimpanzee brain and a human brain are striking. The most obvious difference is that of size. A chimpanzee's brain weighs approximately 450 grams, while a typical human brain weighs approximately 1,300 grams.[22] In terms of bodyweight, there isn't much difference between us and chimpanzees, implying that our brain is about three times the size you'd expect for a typical primate of our body size. This relationship between brain and body size, a measure called encephalization, has long been suspected of holding a key to human intelligence.[23] And indeed it is important. Your brain's first business is to tend to the needs of your body and to survive in the world. Our oversize brain may give us some extra mental breathing space, particularly for the prefrontal cortex, which underlies key elements of our humanity. When it comes to this organ, at least, size does seem to matter. But is it just a matter of size?

Big brains take a long time to grow, a kind of growth that is turning out to be much different from the orthodox view of a decade ago. Back then, as we began working together on the question of how the mind emerges from the developing brain, the received view was that our brains develop very rapidly up to one or two years of age, producing a huge excess of nerve connections, the synapses through which brain cells communicate. After that, it's all downhill. That is, as development continues, synapses supposedly get whittled away a bit like the way a sculptor knocks away excess marble to uncover the figure inside, with this paring-down process underlying cognitive development.[24]

Within the last few years, there has been a dramatic change to this story, although it has been largely confined to specialists working in the field. The development of your brain to full maturity is both far longer and more growth-oriented than once thought. To give you a few main landmarks, your brain reached almost 90 percent of its final size by the time you were about five years old, although it kept growing well into adolescence.[25] In contrast, a chimpanzee reaches this milestone before the age of two. There was something else about brain development that intrigued us. A decade ago, it was thought that different brain regions develop at the same global rate.[26] It is not known that development occurs at different rates in different parts of the brain.[27] The visual area is already about halfway to its mature size when you are born. In contrast, parts further removed from your sensory windows

are much less mature at birth and take far longer to develop. The neurons in some regions of your prefrontal cortex, for example, underwent
a phenomenal amount of growth after you were born, the majority of it
occurring after the age of two.[28] Recent studies using structural brain
imaging confirm that every region of the cortex continues to grow
larger throughout childhood.[29]

Is it a coincidence that the prefrontal cortex, the part supporting
many key human capacities, such as our ability to develop a sense of
self, also appears to have such a protracted development? Late maturation makes no sense from evolutionary psychology's vantage. Imagine
buying a computer preloaded with tons of software only to turn it on
and discover it takes three years to boot up. Wouldn't evolution have
selected for humans whose brains could get online as soon as possible?
After all, there's nothing to be gained from waiting around twiddling
your synapses when there's a world full of genetic competitors to van-
quish. Evolutionary psychologists ignore this issue, suggesting that the
brain's postnatal development is limited.[30] But if the postnatal development of modules is limited, why would they take so long to get
going?

The prolonged vulnerability of infants and the enormous parental
investment humans require sound like a surefire recipe for extinction.
When you imagine your ancestors having to care for their helpless infants for so long, you really wonder how they survived at all. One thing
about evolution seems clear: Risky paths inevitably lead to dead ends
unless there's some real payoff in taking that route. As we worked out
the details of cultural biology, we kept asking ourselves what the payoff was in slowing down and staging our brain's development.

To begin to answer this riddle, we needed some way to grapple with
the brain's complexity. The reductionist approach of traditional science—reducing a system to its smallest components—falls short when
tackling the brain's puzzles, because the brain is a complex system
whose operation depends on the simultaneous interaction of forces,
from the smallest level of single molecules to interactions at the level
of the whole brain.

As an analogy, consider a tornado. To understand how a tornado forms, atmospheric scientists need to combine what goes on at the smallest scale—the interactions among molecules to form drops of water—with what goes on at huge scales, such as global currents of air flow. Predicting tornadoes is incredibly difficult because their formation depends on how all these forces interact in complex ways on a wide range of distance and time scales. Large-scale computer simulations provide hope for understanding complex systems like tornadoes. Their power lies in linking all the different scales and seeing what happens when they interact in the virtual world of the computer.

Brain researchers are like atmospheric scientists confronting a tornado. And the computer is turning out to be a crucial tool here, too. In fact, there's a deeper reason for using computers to explore the brain. We've mentioned that the brain has been likened to a computer. This is more than mere metaphorical talk. Your brain is literally a special type of computer, though very different from the kind sitting on your desktop. Instead of waiting for information to be tapped in on a keyboard, your brain spontaneously takes in information from the world and performs complex neural information processing whose principles are only beginning to be understood. Unlike a desktop computer, your brain has no central processing unit that interacts with memory registers. Instead, it processes information through a vast array of parallel networks, each made up of millions of cells wired together and often distributed widely throughout the brain.

One of the most promising uses of neural networks is in exploring development, and it is through this use that we begin to find the preliminary answers to our questions. Using this approach, we can incorporate the detailed information neuroscientists have learned about real brains into neural network computer simulations. The synapses in neural network models can be programmed to mimic key properties of real synapses, for example. Once we have set up this model, we can study how information from the world coursing through networks can help them grow. In other words, we can assess how activity can guide brain development, how and when it cues the creation of more synapses, when to add a dendritic branch, even when to remove connections. In essence, we can begin to "grow" silicon brains.

This new approach confirms that the old "precisely programmed" view of development was misguided, in part because intuitions that development was a fragile process that required precise orchestration turned out to be erroneous. In its place, a different view is emerging that replaces intuitions with simulations. At the center of a view we refer to as "neural constructivism" is the realization that the brain has powerful ways of using information from the world to build the complex circuits of the mind, a capacity sometimes called "self-organization."[31] From this perspective, brain development is far more intriguing and dynamic than anyone thought.

As we explored the dynamics of brain development, we realized that our findings would undermine a classic distinction between brain growth and learning. This distinction is what makes developmental psychology and developmental neurobiology different disciplines, with psychologists in charge of studying how infants learn and neurobiologists in charge of studying how the brain grows. According to this time-honored distinction, brain growth was a precisely programmed maturation, regulated internally with little environmental dependence, just as legs and arms mature. Because arms and legs mature without instructions from the environment, psychologists don't study how you *learn* how to grow them. Learning, on the other hand, was akin to making software changes. As a matter of principle, it was believed that learning could not change the brain's hardware. Your computer does not grow new circuits as you type, for example. But the idea of self-organization blurs the distinction between software and hardware. The new models show how interacting with the world is a special kind of learning. It actually changes the brain's hardware and helps build it, through a very slow type of learning that computer models showed to be feasible. In so doing, they overturned one of the cardinal distinctions in psychology: The divide between learning and brain maturation no longer made sense.

We called this brain/environment interaction "constructive learning," a powerful type of learning that helps to build the circuits of the mind.[32] Its lesson is that the magnificent cortical structures supporting a human mind are not entirely hardwired by evolution, but are built by genes interacting with the world. Your experience with the world literally helps build your brain, which in turn gives rise to your mind.

We think constructive learning is a better, more powerful way to build a mind than precise programming. Constructive learning makes the evolutionary vulnerabilities of slow development worthwhile. Indeed, we suspect that protracted cortical development maximizes the world's influence on our brain, thereby making being human possible. The longer the brain develops, the more opportunities there are for specific contributions from the world to guide the buildup of more and more complex circuits—opportunities that are essential to build a mind capable of navigating the complexities of human existence.

Something else about the findings on brain development deeply intrigued us. Earlier, we mentioned that the brain's development isn't uniform everywhere. Parts near the sensory window get built earlier than parts farther away. In other words, the brain stages its development in a hierarchical manner.[33] This provided us with a critical hint as to why human development is protracted and how being human might have been made possible. The other critical piece of the puzzle stemmed from reflecting on a central theme of information processing in the brain. The brain appears to process information in a roughly hierarchical manner; that is, brain regions near the sensory windows represent quite basic features of the world and support specific sensory and motor functions. Areas farther away encode more abstract or complex features and have a more integrative function in a hierarchy of information.[34]

The brain's staged, hierarchical development thus mirrors its mature hierarchical organization. This suggested to us a deep connection between constructive learning and hierarchical brain development. If information from the world plays a role in building the brain's circuits, then it would make sense for regions near the sensory windows to be built as the foundation for the subsequent development of higher regions. Making development more protracted would allow for building more complex representations in ever more integrative regions—so long as the environment could play a role.

As an analogy, suppose you are constructing the Empire State

Building. You have to build the foundation before you can start building the first floor. Once the first floor is in place, you can add more and more levels. Finally, you can build the spire at the top. You must follow this sequence because each level is supported by what lies below.[35] Earlier theories of brain development that were based on a uniform development throughout the brain had no room for this staged sequence of development by which each level serves as the foundation for the next. It was no coincidence that these theories minimized the role of the world in development. Hierarchical development and the hierarchical organization of the brain go hand in hand with constructive learning to reveal the benefits of protracted development. You can't build the mind's magnificent structures—language, complex thinking, and social skills—without layer upon layer of experience.

Human brain development, then, is an immensely rich cascade of genetic checkpoints and environmental shaping, designed to build more and more complex stages, each advance depending on the last. Our interactions with the world play an essential role in this process. Even differences in the richness of our environment can mean the addition or loss of literally billions of synapses. Thus, we are born with our eyes and ears open, ready to engage and absorb the world even though we cannot hold up our head or move our body in a controlled way. As these sensory areas develop, they begin shaping more remote parts of the brain, the floors above the sensory foundation. If your brain were built overnight, it would never get the chance to build your powerful and unique mind. You've got to grow for a long time to be smart.

The World Can Make You Free

We started this chapter by asking how the 1.6 percent genetic difference between chimpanzees and humans—a difference that is smaller than that between two kinds of gibbons—could hold the key to making humans. Let's now explore how the new findings provide some clues for how nature found a way to make a human so economically.

Building a human economically is difficult if you think of genes conventionally as blueprints, the sole repository for the brain's development. The first clue that the blueprint metaphor is deeply misleading lies in realizing that there is no simple relationship between an organism's complexity and the size of its genome. Consider, for example, that our brain has a fantastically larger number of neural connections than the brain of our nearest relatives. Our prefrontal cortex is six times larger in absolute size than the chimpanzee's. Yet this huge increase in structural complexity comes without an increase in genome size. Given how closely related we are to the chimpanzee genetically, it seems unlikely that nature could write a human blueprint in the small differences between humans and apes. Our genome is about the same size as the mouse's, or any other mammal's for that matter. In fact, we share a great number of our genes with mice, making them a useful model system for genetics research. Even more curious, some plants have a larger genome than we do.

The mystery deepens when you consider that mammalian brains come in a wide range of sizes, from the shrew brain that weighs .0584 gram to the baleen whale's brain, which is a hundred thousand times larger. On the face of it, the enormous diversity of brains in nature might undermine any attempt to find common themes of brain evolution. Yet one of the seminal—and most surprising—findings in the genetics of development was the discovery of homeobox genes—key developmental genes—and their striking conservation.[36] These discoveries not only revealed remarkable similarities, known as homologies, in the genes underlying the development of mammalian brains but also showed that the developmental programs of insects and mammals, for example, are remarkably alike in that similar, or homologous, regulatory genes have been identified that control the key processes of embryonic brain development.

These results help make sense of the fact that there is little obvious relationship between genome size and brain complexity. Complexity doesn't come from just adding more details to a genetic blueprint, which turns out to be a deeply misleading way to think about how genes work. Instead, an important route to building more complex structures is by using the same genes in different ways: altering how, when, and how long genes are used.[37] This is why insects and humans can share so many of the same developmental genes yet be so strikingly different in appearance and behavior. This understanding suggests a powerful way for evolution to change brain structures: by altering developmental programs, particularly by changing the timing of events during development, a process referred to as "heterochrony." These remarkable insights into the genetics of development have provided the foundation for a new, developmentally centered evolutionary view known as "evolutionary developmental biology."[38]

During hominid evolution, large brains may have come about in part by changes in the timing of developmental events, resulting both in larger brains and in our long postnatal development. A vast array of new, uniquely human genes wouldn't be necessary to make a human. Instead, tweaking the timing of developmental events by altering the function of some key regulatory genes could underlie pronounced alterations in the brain.

Growing evidence for this idea stems from examining how brains develop.[39] Most brain cells arise from a pool of cells called precursor cells, which multiply by continually dividing. The site of this rapid cell multiplication is known as the proliferative zone. After a certain length of time, some cells stop dividing and begin their journey to a distant part of the brain, where they become specialized neurons. One possible way to regulate how large a brain structure becomes lies in regulating the proliferative process, since the size of a brain structure depends in large part on the number of neurons there. Hence, the longer the process of cell proliferation lasts, the larger the brain.[40] It appears that this sort of change may underlie key differences among primate brains.[41]

Extending the duration of cell proliferation would affect the entire brain, which appears to happen in many cases. A different strategy would be to more finely regulate proliferation for each brain region or for groups of regions. This would change a specific structure without having any effect on others. Some argue that specialized structures have been picked in just this way by evolution, such as brain structures used in navigation by foraging species.[42] Such a mechanism could change the relative proportions of brain areas as a new group of species emerges, which occurs on a long timescale, estimated at ten million years.[43]

Thus, these two mechanisms of evolutionary change are not incompatible, but operate on different timescales, one leading to changes within a group and the other leading to a reorganization which in turn leads to a new group, or taxon.[44]

Enlarging the primate neocortex by adding rounds of cortical neurogenesis during evolution has the consequence of making postnatal development more protracted. Constructive learning makes the development of the neocortex increasingly sensitive to the environment. Since there's a limit to how much genes can specify, or how large brains can become, at some time during evolution, making brains more complex demanded escaping this limit. The escape route came by finding a partner to hold part of the brain's developmental program— probably relatively simple parts of the natural world early in evolution. Over time, the developmental program came to rely more and more on the world. The process of increasingly incorporating the world into the brain's developmental program, a process we termed *progressive externalization,* is a way to make the brain more adaptable in an uncertain world.

The brain's developmental program, then, is not simply contained in the genes. Rather, it is held implicitly in the interaction between genes and the world. And it takes a special kind of interaction between the brain and the world to build a mind, an interaction that we outlined earlier in our theory of constructive learning. The structure of the world holds a great deal of enduring information, making it unnecessary to encode this information somehow in genes. But partnership with the world also has a price. Consider this: Many animals have enzymes that can make vitamin C from other foods. At some point in their past, humans had access to fruit, which is rich in vitamin C. Because we did not need to synthesize vitamin C internally, our bodies lost this enzyme, and we now run the risk of developing scurvy if we don't obtain vitamin C through our diet.[45] Satisfying your bodily need for vitamin C has been transferred into the environment, making you dependent on that environment. Thus it is in myriad more complex ways with the brain.[46]

Beyond Swiss Army Knives

Swiss Army knives are wonderful inventions. But their specialization comes at a price, as they aren't much good for anything beyond their preordained functions. As a metaphor for the brain, as we explained in chapter 2, they are deeply misleading as they obscure the brain's robust flexibility. Consider a landmark experiment in the frog to see how striking this robust flexibility can be.

Having eyes in the front of your head gives you binocular vision: The images seen by each eye overlap, allowing your brain to calculate depth and other essential information. This ability, along with other cues your brain culls from the world, enables you to see the world in 3-D. When the information from your eyes is relayed into the cortex, the input is segregated into bands of alternating left- and right-eye input. When an anatomical dye is injected into the optic nerve, it diffuses upstream to the visual cortex, staining these alternating bands, called ocular dominance columns, which resemble zebra stripes.

The downside to having binocular vision is that your peripheral vision is reduced. If you had eyes in the back of your head, you would be able to see what's sneaking up on you. The frog doesn't quite have eyes in the back of its head, but it does have eyes on the sides of its head, like two radar screens, to track incoming flies for lunch. Now, here's the intriguing part. What would happen if you transplanted a third eye right in the middle of a developing frog's head? Totally unexpectedly, the frog might have binocular vision. Martha Constantine-Paton and her colleagues at Yale performed this experiment.[47] When they looked inside their three-eyed frog's brain, they found that it had made the stripes found in binocular animals. It seems unlikely that the frog had a genetic plan for the rainy day when a third eye might magically appear in the middle of its head. Instead, the frog had the same mechanisms that binocular animals have, but used them for different purposes. When information came in from a new eye, however, these mechanisms were then used to make stripes. In other words, the frog's brain was able to accommodate novel sensory inputs. Such flexible developmental mechanisms interacting with the world make turn-on-a-dime survival changes possible. This flexibility is called evolvability. The human capacity for constructive learning and the process of progressive externalization demonstrate how the pressures of evolution have resulted in flexible developmental mechanisms, not hardwired, precisely programmed mental circuits.[48]

Recent progress in brain research also has revealed that brain development isn't the fragile, highly improbable process it was once thought to be. Apparently sloppy-looking strategies can build structures of great complexity. The process is robust, flexible, and opportunistic. These themes suggest that evolution is not a blind watchmaker because our brains aren't at all like watches.[49] We now realize that a flexible brain develops through a long and complex interaction with the world; a special brain-world interaction builds the circuits of the mind. As we reflected on these new lessons, we realized that the starting point for rethinking the nature of human nature lay in the following key insight: With humans, some new process was found that profoundly increased the complexity of the brain-world interaction that builds the mind.

Culture has many levels, from social transmission of behaviors such as tool use observed in nonhuman primates to symbol and language, rituals, and institutions, which are highly developed in humans. The Canadian psychologist Merlin Donald refers to human culture as a "cognitive web" and suggests that humans found a way to build complex cultures that free the mind from its individual isolation to partake in a widely distributed system of knowledge shared across many nervous systems.[50] This makes human mental life irreducibly social. The comparative primatologist Michael Tomasello suggests that humans have a species-specific type of social learning that allows us to form a sense of self and understand others as agents with intentions that are similar to our own.[51] Human development critically depends on the ability to understand others as persons with minds, full of beliefs, desires, hopes, and other intentions.[52] This ability is the entryway into language, which further opens new levels of social interaction. In recent years, there has been a great deal of speculation that autism involves a disruption in this social understanding.

It is deeply intriguing to note that social understanding depends on the prefrontal cortex, which, as we mentioned earlier, has regions in the human brain that are intimately connected to self-awareness and social cognition. The development of the prefrontal cortex, among the most protracted in the brain, is sensitive to the social environment it is immersed in. In this way culture exerts a central influence on the development of the prefrontal cortex, helping to form the social cognitive skills necessary for the complex social life we share. Earlier, we noted that the brain's development is staged, with some regions developing later than others. Culture reflects this staging model, too. Human development is like a set of nested Russian dolls.[53] The first social exchange that the child experiences is with the caregiver. Toward the end of the first year of life, the child begins to draw external objects into its interactions with caregivers.[54] Through these social interactions, language emerges, and the child's world continues to expand, to the world of child care and school, peer groups, neighborhoods, and eventually to the larger society.

This journey strikes us as a beautiful example of synchrony between the developing brain and the world it is immersed in, each building on the other, the child's cultural world expanding in lockstep fashion with the developing brain. Culture, then, contains part of the developmental program that works with genes to build the brain that underlies who you are. Thus it seemed to us that "cultural biology" was the most fitting name for this view, which sees these two forces as inextricably intertwined.

And with this complex interaction, something very special followed: A large, long-developing brain could be more responsive to the world around it, capable of adapting quickly to changes in the cultural world, accelerating the timescale of change from a slow, evolutionary one to a fast developmental one that changes in historical time. The loop is thus closed: Culture helps to shape your brain, which in turn creates culture, which acts again on the brain, reverberating throughout the generations. We are still only beginning to appreciate how this immensely rich feedback loop leads to much of what we recognize as human.

Culture and Biology

In this chapter we've seen the power of constructive learning, of extending and staging the brain's development to allow the brain to be built through rich interactions with the world. And we've seen an elegant solution to the question of how our brain seems to have bootstrapped itself into humanity, despite its many similarities to the brains of other species. The most basic lesson is that human evolution is not simply the accumulation of more and more specialized brain functions. It is the advent of robust developmental programs, made possible by increasingly complex interactions with the world. It's not a question of nature or nurture, but of an intersection of the two that is so thorough that it dissolves that simplistic dichotomy. The research we've highlighted compels us to see the brain and the world as an integrated system, a fabulously tangled web of interactions.

Flexibility and responsiveness to the environment are quite different ideas from those of the modern image discussed in earlier chapters. They are the emerging themes in our new understanding of human brain development and evolution itself. One crucial caveat: The emerging view does not mean that our brains are general-purpose computers or mounds of shapeless clay the world kneads into shape. Nor does our capacity for culture and the behavioral flexibility it bestows signify that we are unfettered from our biology. Our emerging understanding reveals why these capacities are a reflection, not a refutation, of our biology. Indeed, in later chapters we will look within ancient brain structures that guide the process of prefrontal development and constrain who we can become.

Why did our brain become the way it is? What forces acted on our ancestors' brains to confer the human brain with its specific capacities? In the next chapter we'll discover that our ancestral past was much different from what has been supposed—a difference that holds the key to understanding who you are and how you came to be.

Life on the 4 Far Shore

Crossing the Mental

Rubicon

n the stillness of the dawn, the figures of six girls can just be made out from the shadows. For the last twenty-four hours these children of Uganda's Sebei tribe have lived the secret rites marking a Sebei girl's passage into adulthood. Now, as the sun sits just below the horizon, they prepare for their final rite: the female circumcision ritual. As they lie down in a row for the painful operation, elders appear out of the shadows to crowd solemnly around. All eyes turn to the ritual circumciser, who raises a hand-forged iron knife high into the air. Down it falls as he begins the first procedure on the most senior girl. To the shock of onlookers from the West, the girls of the Sebei tribe press their parents to allow them to take part. But with the first cut one girl stands up and refuses to allow the mutilation to go any further. The crowd, silent a moment ago, is shocked by her behavior. They begin to taunt her, hurling insults and sticks. Ashamed, she runs off, heading for a tree the Sebei reserve as a place of suicide. Suddenly transformed into an outcast, stripped of who she is, she now seeks the comfort of death. But as she scrambles up the tree, a group of men catch hold of her dangling foot and begin to drag her down. As she finally falls out of the tree, they pin her down to complete the operation, her screams piercing the silence of the dawn.[1]

Here in this terrifying scene are behaviors unlike any witnessed in the rest of the animal world. Why was this young woman initially so willing to undergo a horrific, disfiguring operation? When she flinches, and then flees in shame, a uniquely human emotion,[2] why does suicide appear such a ready choice? For a woman just approaching her reproductive years, seeking suicide seems a deep violation of natural impulses. And why would a group of men be so hell-bent on enforcing their society's customs?

Shocking rituals like female circumcision stretch our capacity for cultural sensitivity, but cultures throughout the ages have surrounded life's passages with rites that leave deep scars. If you've seen the video footage showing marines pinning recruits right through their skins as the recruits writhed in obvious pain, you know that even so-called modern society hasn't abandoned painful rites of passage. Every society is full of deeply symbolic customs, often ones that you willingly take part in despite a high toll. These rituals form part of your culture, creating a moral order, a web of significance that makes you who you are.[3]

Culture is often regarded as the unique province of humans. Almost from the moment that seventeenth-century philosopher René Descartes drew a hard line between our mental life and the mechanical operation of beasts, however, comparative researchers have provided pointed critiques of this divide. For decades, primatologists have searched for signs of culture in apes and have discovered many mental subtleties once thought beyond the capacity of nonhuman primates.[4] Later in this chapter we'll dissect the notion of culture into various components, some of which we share with nonhuman primates. As we noted in the previous chapter, though, humans are capable of some cultural processes that appear to be unique to our species.[5] In particular, these appear to depend on a social understanding of others that facilitates a particular type of cultural learning, which in turn allows us to construct complex social worlds. The rapid pace of human historical development suggests that this capacity has powerful implications. Thus, it is reasonable to suppose that some sort of mental Rubicon, some important difference in how our minds work, separates us from our closest relatives.[6] Humans live on the far shore of meaning and significance, symbol and ritual, creatures of both biology and culture.

Reflecting on this, we realized that our first step in understanding cultural biology's lessons would be to retrace the steps our ancestors took as they crossed the mental Rubicon to become us. What forces propelled them into this world of human culture, and what lessons do these first moments of humanity hold for who we are?

From La Jolla to Africa

In 1996, a group of researchers from a wide range of backgrounds formed the La Jolla Initiative for Explaining the Origin of Humans and began holding regular workshops at the Salk Institute.[7] The group drew from traditional methods in archaeology, which depend on the chance discovery and preservation of fossils, and from recent techniques in molecular genetics that offer new clues to our origins. In addition, advances in allied disciplines, such as paleogeology and paleoanthropology, are giving us a different picture of what conditions were like for early humans. Finally, neuroscience is making it possible to search for changes in brains that enabled our ancestors to journey across the mental Rubicon. As members of this group, we benefited from far-ranging discussions on what is known in each of these disciplines. Our thinking on the origin of humans has been strongly influenced by this group.

We cannot begin to understand who we are without knowing where we came from. Intertwined with a story of our origins is also one recounting why we now live as we do, even perhaps giving us a vision of the future toward which our human character will lead us. For try as we might, we cannot escape the history that is buried inside of us.

Although our understanding of the distant past is fragmentary and incomplete, there are some tantalizing clues. So let's go back to those earliest moments when our ancestors first distinguished themselves from other apes. Somewhere back in time, our ancestors literally invented themselves, and so invented us.

A Babe in Toyland

The search for our origins runs into an obstacle right at the outset: the unfathomable time line of history. One of the greatest obstacles facing Darwin's ideas was the sheer expanse of time that they implied: His theory flew in the face of the general consensus that the earth was about six thousand years old, a date determined by biblical scholarship. In the end, it meant adding about 4.6 billion years to the earth's age and recalibrating all other events from there. Life first emerged around 3.5 billion years ago, in the form of single-celled creatures that shared the earth for almost 3 billion years. Only in the last 600 million years have multicellular plants and animals existed. From the first signs of multicellular life, it took 550 million years for dinosaurs to appear and head the food chain, around 200 million years ago. Some 60 million years elapsed between the extinction of the dinosaurs and the rise of anything looking faintly human.

Evolution is like a branched tree, with living creatures at the tips of the branches. Humans have not descended from chimpanzees and gorillas; rather, they are our cousins, since we share common ancestors. Scientists once thought that humans and apes split from a common ancestor between 15 and 30 million years ago, with the great apes evolving in one direction, subsequently becoming subdivided into orangutans, gorillas, and chimpanzees, while we went off in another direction. Now that scientists are able to use genetic differences as a kind of molecular clock, a new picture is emerging. The orangutan has always seemed a bit of a misfit in the accepted scenario, and genetic measures reveal that it differs from us in about 3.6 percent of its DNA. Orangutans departed the evolutionary bus that led to humans between 12 and 26 million years ago. Gorillas differ from us in about 2.3 percent of their DNA, having exited the bus around 10 million years ago. The 1.6 percent difference between ourselves and chimpanzees noted in the previous chapter means that the last fork in the road came only 6 million years ago.

In evolutionary terms, 6 million years is a mere eye blink. And although no common ancestor has been found among the fossil record yet, the oldest hominid specimens found so far lived in Ethiopia around 4.4 million years ago.[8] These four-foot-tall creatures, known as *Ardipithecus ramidus,* appear to have lived in forests, not on the open savanna our ancestors are so closely associated with. It is questionable whether *A. ramidus* walked upright. By around 4.2 million years ago, however, *Australopithecus anamensis* appeared in Kenya and was walking upright. A variety of *Australopithecus* has strutted upon this stage. The most famous is Lucy, discovered by Donald Johanson and colleagues in Ethiopia in 1974, who got her name sometime during a night of celebrating her discovery while "Lucy in the Sky with Diamonds" was playing. Lucy lived around 3.18 million years ago and walked upright, but still had a brain not much bigger than a chimpanzee's. Her descendants dominated the scene until 2 million years ago. Then, a newcomer appeared.

This newcomer, the first of the *Homo* lineage, was named *Homo habilis,* literally the handy man. The brains of this species were about 50 percent bigger than Lucy's, perhaps explaining why stone tools were found among their remains. A brief two hundred thousand years later, another hominid appeared, *Homo erectus,* arising first in Africa and then dispersing into Asia. By around four hundred thousand years ago, the descendants of *Homo erectus,* archaic *Homo sapiens,* could be found in Africa and Asia. Their brains were even larger, weighing between 1,100 and 1,400 grams. However, anatomically modern humans didn't appear until sometime between one hundred thousand and two hundred thousand years ago, and their brains weighed in at between 1,200 and 1,700 grams.

Fossils speak to us only indirectly and are mute when it comes to revealing how our ancestors lived. Anthropologists thus face the daunt-
ing task of reconstructing a past that offers few clues to the real dramas
that must have taken place. One of the clues anthropologists have
about the mental life of ancient peoples are the tools they left behind.
With the appearance of anatomically modern humans, there are some
tantalizing signs that life was changing. But even as recently as one
hundred thousand years ago, the tools our ancestors used appear little
changed from those used much earlier. Indeed, the unequivocal artifacts of humanity would begin to appear only fifty thousand years ago.
In a deep sense, you are a babe in Toyland, the tenderfoot in an ancient
drama.

Signs of Human Life

Among all the things that mark us as human there is perhaps nothing
more distinct than the human reaction to death. In Alejo Carpentier's
haunting novel *The Lost Steps,* his characters search for the origin of
music in the primordial Amazon rain forests. Their hunch that music
stems from the imitation of animal calls is shattered when they come
across a burial ceremony among ancient people and realize that music
imitates their wailing grief for the dead. While Carpentier's theory may
have been fanciful, it makes us wonder how our ancestors dealt with
death. When did death become a human experience?[9] Some clues
emerge at an ancient burial site near Sungir, Russia. There, two youngsters were buried some twenty-eight thousand years ago. Those who
buried them adorned each youngster with clothing into which three
thousand ivory beads had been sewn. Modern estimates suggest that
each bead took about an hour to make. The youngsters are also
adorned with carved bracelets, pendants, and necklaces. Two six-foot-
long mammoth tusks are buried alongside them.

This elaborate burial reveals many things about these people. The
inclusion of possessions in the grave suggests they believed in an af-
terlife in which the goods would be useful to the two youngsters. It

also suggests that the children's burial was accompanied by ritual and that their life, too, was full of symbol and significance. Clearly, these humans were on the far shore of the mental Rubicon, inhabiting a world of meaning.

We don't know when our ancestors crossed this mental Rubicon. The archaeological record reveals that around forty thousand years ago human sites in Europe were suddenly being filled with bone and ivory carvings, bone needles for sewing, beads and other personal adornments, and ritual burials. Based on the sudden appearance of these items, certain anthropologists suggest that some spark of modernity appeared around that time, igniting human invention. Since these events took place over sixty thousand years after humans spread out of Africa, though, the explanation is unlikely to be a biological change.[10] Others argue that similar events were occurring independently in Australia and China and that there was no sudden explosion of human invention.

If we can't be certain about when our ancestors crossed the mental Rubicon, perhaps we can at least begin to understand what forces drove them to create the human culture that placed them firmly on the far shore and how these forces shaped our nature.

The artifacts of human culture are unlike anything ever seen in the three-billion-year history of life on earth. Consider, for example, the haunting painted images in the caves of Lascaux. Expecting simple stick figures, we see instead graceful, flowing images that today would be the mark of an artist. Indeed, it is their beauty, the sense that their creator had captured the essence of these animals, that leaves us with a feeling of connection with this artist who lived so long ago. This we understand as the work of human hands and mind, of someone whose life we would recognize as human. Gazing at these images, we realize that their creator must have had a profound sense of self. And we realize that in the act of painting, the artist likely had a deep urge to make life meaningful and to communicate that meaning to others. These frozen images are the primordial answers to the question of who we are. From these roots, the human story unfolds across the centuries, captured in myth and art, the generations connected by their common search for identity and the desire to communicate their story.

Painting the events that hold deep significance for one's life is the sort of behavior recognized as distinctly human, akin to the behavior that underlies the disturbing story of the Sebei girl that opened this chapter. As some put it, humans have finally invented humanity.[11] Their world no longer consists of just their physical environment but embraces a symbolic level that gives meaning to the past and to the future.

With the rise of human culture, an interesting thing happened. The pace of change broke loose from the geological time line and accelerated exponentially,[12] as cultures expanded. Societies became increasingly complex, adding navigational skills, food gathering and preparation, ritual, and decoration to their cultural repertoire. As the last Ice Age retreated about twelve thousand years ago, humans made the trek from Siberia to North America, descending into the Americas. The domestication of farm animals and the rise of agriculture occurred ten thousand years ago. Towns began to appear.[13] Five thousand years ago cities, with their specialization of labor, government, and social stratification, arose in Mesopotamia, modern-day Iraq. Writing, first pictographic and then more abstract, emerged, as did the calendar and mathematics. Slavery appeared, along with modern warfare. Recorded history began its furious march into the present.

We were interested in what forces might have caused human culture to emerge and, specifically, how those forces acted on the brain to give rise to human culture.

The Search for Eden

All mythical accounts of human origins intertwine human nature with the worlds our ancestors inhabited. In modern evolutionary theory, the natural environment's influence on animal behavior is equally strong. Indeed, evolutionary biologists typically regard an animal's nature as a highly specialized adaptation to environmental challenges. Consider Darwin's reaction to the thirteen species of finches among the Galápagos Islands, a series of volcanic islands six hundred miles off the coast of Ecuador. Since the islands have never been attached to the South American continent, all species of plant and animal had to travel six hundred miles over ocean to get there. The most striking difference among these thirteen species of finches was their beaks. The only apparent reason why there would be so many different species of finches among the Galápagos Islands was if the differences among them reflected different environmental challenges. Since beaks are used to gather foodstuffs, the match between different kinds of food sources and beak structure proved to be a pivotal insight in thinking about the origin of species. According to this view, evolution adapts species to fit the needs of their environment. Are you, too, bundles of specializations that reflect the challenges your ancestors faced in the worlds they passed through on their way to becoming you?

In the 1970s, E. O. Wilson argued that the human mind is as highly adapted to modern society as Darwin's finches are to their world. Wilson's speculative view implied that any alteration from the status quo was a violation of nature: If people live in households in which the majority of men work outside the home while women care for the young, they do so because these behaviors reflect evolutionary specializations as surely as do pointed beaks.

More recently, evolutionary psychologists have proposed an alternative view, suggesting that our origins lie in a long, enduring way of life over eons of evolution. According to this theory, 95 percent of our evolutionary past was spent in a stable way of life as hunter-gatherers on distant savannas. As science writer Matt Ridley puts it, "for more than a million years people lived in a way that couldn't have changed much."[14] The idea of a stable and enduring way of life is crucial to this account of our origins. It means that each generation will confront more or less the same problems, allowing our minds to adapt to enduring problems gradually over the ages, perfected by evolution the way a watchmaker tinkers with the cogs of a watch. The finch gets a pointed beak; we get specialized mental modules for navigating the world of a hunter-gatherer on the savanna.

The implication of this influential view is that we are misfits in our own age, equipped with minds belonging to the Stone Age, not the

Computer Age. Many of the foibles and indeed much of the misery of modern life are explained as stemming from the mismatch between a mind specialized for an ancient, lost way of life and the demands of our complex postindustrial life. Thus, murder, rape, and other once-adaptive strategies are wired into our brains despite the misery they often bring when employed in our society.[15]

This story of our origins has the ring of the fall from Eden. It regards our modern life as a fall from a distant past when our mental adaptations were in tune with the world. But this fall begs for explanation. What is evolutionary psychology's apple? If our minds were so adapted to our ancestral environment, the way a honeybee is adapted to nectar gathering, why did we leave that haven? Why did the world of culture and a mind smart enough to occasionally fool Mother Nature ever emerge?

The first step in solving this mystery is to see that evolution thrives on pressure. Think of the auto industry. For most of the past century, American manufacturers only had themselves to compete with in the U.S. market. By sheer market share, General Motors, whose sales surpassed the entire GNP of Italy during the 1950s, could simply dictate customer tastes. Other U.S. automakers were happy to follow along. The result was little innovation in the industry. But when Japanese automakers introduced competitively priced cars of superior quality in the 1970s, the balance was shattered. Suddenly, American cars were less fit, and more than one U.S. auto manufacturer was staring extinction in the face. Either you improved or you were pushed aside.

Evolution needs these kinds of pressures. Without them, you're left unchanged or drifting without direction. What pressures on our ancestors finally pushed them out of their Eden? One candidate is the environment. Yet evolutionary psychologists think that the ancestral environment had not changed much, offering little pressure. Finding the answer to this riddle would lead us to rethink the basic picture of our past and the forces that spurred our origins.

How to Win Friends and Influence Primates

Animals pitted against other animals for survival create evolutionary pressures. Think of the world as a marketplace shared by many species, all looking for market share, but in this marketplace you can literally eat your competitor. In other words, the leopard gets its spots that allow it to sneak up on the zebra, whose stripes help conceal it from the leopard, and so on. Another analogy used by evolutionary theorists, especially during the cold war, was that of nations in an arms race. But sometimes the enemy comes from within: The real race may not be among different species, corporations, or nations, but among members of the same group trying to outmaneuver one another in the genetic arms race. In this scenario, since your ancestors inhabited a setting that didn't change much, and they didn't have to work too hard eking out a living, they turned their competitive energies on each other.

Evolutionary psychologists don't think the world put the screws to our ancestors; their friends did. As a poet has phrased it, we have seen the enemy and he is us. Robert Wright puts it this way: "[V]arious members of a Stone Age society were each other's rivals in the contest to fill the next generation with genes. What's more, they were each other's tools in that contest."[16] According to this view, cultures and the smart thinking they require emerged from this mental arms race, shaping our brain to contain specialized circuits for social warfare, eventually making minds that could occasionally outsmart their maker, evolution.

There's no question that the challenges of navigating a social world helped spur our brain's evolution. In fact, this dynamic will be a central part of our story. But we don't believe that a fall from an evolutionary Eden is the right story of our origins. Other forces acted on our ancestors, shaping their brains—and yours—in ways very different from the evolutionary psychology account.

A Tale of Two Apes

We've mentioned evolutionary psychology's scenario that a common set of problems challenged our ancestors over a million-year span. Why a common set? The idea behind this conjecture is that a species' environment goes a long way toward determining how that species will organize itself socially. Social problems stem from this organization. In particular, how food, sex, and other necessities of life are distributed is key to how a group will organize. So a stable environment means a stable way of organizing, which means an enduring set of social problems, which means evolution can design specialized brains to solve those problems.

Earlier, we introduced you to the bonobos at the San Diego Zoo. As the bonobo world slowly comes into focus, researchers are discovering a remarkably different way of life from that of the chimpanzee.[17]

Chimpanzees live in male-dominated groups that fragment during the day as members search for food. Males take part in what many field researchers describe as chimpanzee politics. There's a dominant male, the famous alpha male, but he doesn't hold his position through force alone. Instead, he forms alliances with other males, in a kind of "friends of alpha" network. He repays their loyalty by sharing food and granting them access to sexually receptive females. Other males may form other alliances in an effort to unseat the dominant male—akin to political coups. For the female chimpanzee, this male bonding can be stressful. All females are subordinate to all males and must give every male ground when he passes by, or face his wrath. For the few days a month when she is sexually receptive, signaled to everyone by genital swelling, a female is carefully guarded by a dominant male, who, as we've noted, may allow other males access as a way to maintain alliances with them. For the female, it's a time of great stress, often accompanied by physical abuse.[18] From the moment she gives birth, she spends her time away from the group to protect her infant from aggressive males. A male might kill her infant, as that would allow her to become sexually receptive sooner than if she continued nursing someone else's infant.

Girl Power, Bonobo-Style

Bonobos are less aggressive than chimpanzees, both within and between groups.[19] In marked contrast to chimpanzees, females dominate
in bonobo groups, and the strongest bonds in the group are among females. Females will rescue other females if a male gets out of hand,
but bonobo males are less inclined to be aggressive toward females
than are chimpanzees. What makes this surprising is that the females
of a group aren't related. In many primate groups, members of one sex
leave when they reach adolescence to join a new group—nature's way
of avoiding inbreeding. As a general rule, strong bonds develop among
whichever sex stays in the group. At about eight years of age, though,
female bonobos wander off to join a new group. The mystery of how
bonobo females form tight bonds was uncovered when researchers
watched how females join a new group. Typically, the newcomer sits at
the periphery of the group and tries to make friends with another female of the group. If the resident female is interested, she'll signal to
the newcomer in a not-so subtle way: She'll lie on her back and spread
her thighs. The newcomer quickly embraces her, and the two females
have sex together in a romping bout of missionary-style genital rub-
bing. Now they are friends.

Bonobos behave like sixties hippies who adopted "Make love not
war" as their slogan. About 75 percent of their sex life has nothing to
do with reproduction. They use sex to smooth over rifts, ease tension
surrounding food, and form bonds of friendship. As the primatologist
Frans de Waal nicely puts it, "chimpanzees resolve sex issues with
power, bonobos resolve power issues with sex." If you visit these remarkable apes with young children, be prepared to explain "the birds,
the bees, and the monkeys." Frans de Waal also notes: "[T]he chimpanzee sex life is rather plain and boring; bonobos act as if they had
read the Kama Sutra."[20] Bonobos engage in virtually all sex acts, from
missionary-style copulation to oral sex (no one knows whether bonobos
consider it sexual relations) in all combinations: male-female, female-
female, male-male, adult-child, and even masturbation.

A key to the bonobo's sexuality lies in the fact that whereas female chimpanzees are sexually receptive for only 5 percent of their adult life, bonobo females are sexually receptive more than half of their adult life, much longer than they are ovulating. A major source of tension in chimpanzee groups revolves around the relatively rare access to sexually receptive females. Sex is much easier to get if you're a bonobo male. Thus, male bonobos, while still forming a hierarchy, don't show as intense competition as male chimpanzees, probably because the rewards of competition aren't as high. They don't seem to care as much who's the boss and don't form the strong bonds of male chimpanzees. With all the love children around, males don't know who is the father of whom, so they don't engage in infanticide. The fact that bonobos do not hunt is also likely involved in the lack of strong male bonds.

Clearly, being a bonobo is very different from being a chimpanzee. What makes these differences even more interesting is that bonobos and chimpanzees split from a common relative between 1.5 and 3 million years ago, making them very close genetically. Yet they live in two different social worlds, each species with its own strategies for navigating those worlds. What made such closely related cousins take such different paths? Could the answer hold any lessons for our origins?

The answer stems mainly from differences in the species' respective ranges. Both live in the great rain forest along the Zaire River, formerly known as the Congo, a rain forest second in size only to the Amazon rain forest. Bonobos live on the southern side of the river, while chimpanzees live on the northern side. Since apes don't swim, the Zaire River acts as an effective border. Both species eat the fruits adorning the trees of the area, but bonobos also eat the protein-rich buds and stems of herbs along the forest floor. Chimpanzees don't have these as a major component of their diet, perhaps because chimpanzees share their range with gorillas, who eat these plants in huge quantities. Harvard anthropologist Richard Wrangham suggests that gorillas may have once inhabited the southern bank, but were pushed out by extended drought sometime after the chimpanzee-bonobo evolutionary split. Gorillas, who are dependent on these herbs, would have retreated into the mountain forests, which are confined to the northern bank. This would have left ancestral bonobos alone on the south side with a richer food source.

This richer food source allows for tighter bonobo group cohesion, affording females more time together, leading to the bonds that would reverberate throughout their way of life and would shape their sexual nature. Today, these ecological differences are reflected in the bonobos' more gregarious nature. Chimpanzee groups as a whole are less cohesive because they disperse more during the day in search of food. In particular, females must go off on their own or with their dependent infants to forage. This leaves them little chance to form bonds with other females. In sharp contrast, bonobos rarely move alone. At the center of their group are females, who spend time closer to each other than males do. Bonobos also call to one another when it begins to get dark, and they shack up for the night together on beds they make high in the trees.

There's another interesting difference. We mentioned that bonobos get much of their protein requirements from herbs along the forest floor, while chimpanzees must find another protein source. Chimpanzees meet this requirement by hunting monkeys, typically in male hunting parties. They surround a monkey or group of monkeys, with some chimpanzees attacking while others lie in wait at the escape routes. This provides another impetus for male bonding and cooperation. In contrast, when bonobos encounter monkeys, they often treat them like play dolls.

The striking differences between bonobos and chimpanzees highlight how strongly the bare necessities of life shape the evolution of social organization, social problems, and group behavior. There's a crucial lesson in this for understanding who we are. You can't separate the question of who we are from the worlds our ancestors passed through on their way to becoming us. If the assumption that the human brain took shape in a stable and enduring world is not true, then there isn't a single set of social challenges that sculpted your brain into a set of mechanisms for dealing with a monolithic world. Earlier, we mentioned that compared with that of your ape cousins, your brain is about three times larger than what would be expected for a typical ape body. Our ancestors' brains were getting bigger by around 2.5 million years ago. But so were their bodies. Virtually all of the human brain's expansion beyond its corresponding body size occurred within the last million years, and mostly within the last six hundred thousand years.[21] Since this final expansion gave our ancestors special powers to cross the mental Rubicon, and since we've seen the degree to which external circumstances can shape brains and the behavior they produce, this led us to the crux of the riddle: In what kind of world did our brain's expansion take place?

There's Nothing Glacial About Glacial Change

The search for the distant worlds that shaped us would be impossible were it not for the fact that climates leave distinct chemical signatures. Through painstaking analysis of ice cores, deep ocean cores, and land and lake sediments, climate scientists are piecing together a surprising history.[22]

Our tendency is to think of climatic change as plodding and incremental, occurring over hundreds or thousands of years. Recent findings, however, reveal a far more dynamic and tempestuous earth. There's a growing recognition that most climatic changes occur in sudden jumps that take from a few centuries to decades, or even just a few years, to complete. Climate change is more like throwing a switch than sliding up or down a hill. What's more, climate scientists are discovering events within events: Within periods of long-term climate change are many shorter-term climatic jumps, defying the very notion of a stable world.

This new research reveals that the last million years was a time of jarring climatic changes, the greatest period of climatic fluctuation since Lucy walked the planet 3.18 million years ago, and perhaps the greatest such period ever registered on the planet. Often within the span of a decade, climates underwent dramatic alterations, from rain forest to arid savanna to steppe.

We realized that this jarring history would radically alter the standard account of our origins. Far from indicating a species born of an enduring set of challenges in an unchanging world, these new clues suggest that we are born of change itself. If this were so, then it would have a profound implication for an understanding of who we are.

One reason for the climate's dynamism lies in the fast and dramatic changes wrought by the ocean's currents. These interact with a host of other forces, including the earth's cyclical circuits of the sun and even the wobble in our planet's tilt.

To get a feel for the complex and tempestuous history of the earth's climate—and how it shaped us—let's start with a visit to the earth some fifty million years ago.

Touching down, you are in a strange, warm, and humid world. At the South Pole you discover no ice caps. Instead, to your surprise, there are trees growing in Antarctica. Nor are there even ice caps in the Arctic. In fact, as you explore Ellesmere Island, you find alligators lurking. You are surprised to discover that warm and humid forests are spread virtually from pole to pole. Stopping in what would someday be England, you find yourself in a tropical rain forest resembling Southeast Asia today. During a later warm period, hippopotamuses roam what would eventually become the busy streets of London.

As inviting as this lush, balmy world is, it is about to end. A precipitous decline in temperature is about to reshape the planet. It will mark the beginning of a long deterioration in our earth's climate that still occurs today. Ice, desert, and grasslands will push back warm forests. A cooling period begins, as ice creeps over the treed landscape of Antarctica. In North America, the typical temperature falls by some fifty degrees Fahrenheit. A brief respite comes at twenty million years ago, but by sixteen million years ago, a second cooling trend pushes temperatures even lower. Antarctica is now entombed in ice. Again, there is a respite some five million years ago as temperatures increase, the oceans warm, and trees prosper in the far north. This ends about three million years ago, when another cooling period emerges, one that still grips us today. For the first time, large icebergs form in the northern oceans. A new dynamic is introduced to the earth's climate. With ice crawling over more and more of the earth's surface, the sun's rays are increasingly reflected into space and the earth grows yet colder. Ocean levels drop as much as four hundred feet, redrawing land boundaries and opening new paths across lands previously isolated by water. With ever-growing amounts of the earth's water captured by polar ice caps, the land becomes more arid. In Africa, arid grasslands now push against wet forests, beating them back, with increasing undulations between them. It is a time of both great extinctions and the emergence of new life as life and earth seek a new balance, however fragile and temporary.

Moving forward to the last million years, a time commonly known as the Ice Age, intense climatic instability wracks the earth. A variety of factors, including the previous rise of highlands like the Tibetan plateau, increased temperature differences between equator and poles, and the worldwide decline of forests, all conspire to make climatic disturbances greater than in any previous era. Glacial periods are the great themes now, with massive ice sheets enveloping the northern continents. Warming spells, together forming hundred-thousand-year cycles, interrupt these glacial times. Within these great themes are smaller dramas: short, erratic vacillations as brief as a century. Despite their brevity, evidence from many sources suggests that these vacillations reshape entire habitats with frightening speed.

Now 130,000 years in the past, a glacial period has recently ended, and you find yourself on a planet somewhat warmer and moister than our present world. Moving ahead to 110,000 years ago, it will take less than 400 years for this warm world to end, throwing us back again into a glacial world. Now the temperature jumps back and forth during an unstable period, leaning toward colder and colder temperatures. When northern snows fail to melt under colder summer skies, large ice sheets begin to form. Eventually they become thousands of feet thick. Their crushing weight pushes them outward, scarring the earth, etching great valleys like those of Yosemite. The ice on Antarctica builds to such depths that its weight deforms the planet itself, giving it a slight pear shape. So deeply condensed does the ice become that if you filled your freezer with ice taken from a depth of ten thousand feet in Antarctica it would weigh some three hundred tons.

Over the next hundred thousand years, twenty-four warm periods will interrupt this glacial world. Each change will take a few decades or less and will last from a few decades to nearly two thousand years. Serving as counterpoints to these warming events, there are periods of extreme cold, whose onset seems to be as sudden as the warm periods, and they, too, radically reshape habitats.

Around eighteen thousand years ago, the glacial world is at its zenith. Ice sheets extend into the United States, entomb Great Britain, and cover the major highlands. Arctic tundra blankets France and stretches into Italy. Then, approximately fourteen thousand years ago, the world escapes its glacial grip. Glaciers begin to retreat as temperatures warm to around their present levels. But just as quickly as the earth warms up, it is thrown back into glacial temperatures. By some estimates, in less than five years the average temperature plunged by twenty-seven degrees Fahrenheit, and the effects were felt throughout the globe. This cold phase, known as the Younger Dryas, lasts for thirteen hundred years. And then, as abruptly as it began, it ends. The earth warms.

Humans by now have spread nearly throughout the globe. Agriculture begins almost immediately, and soon our earliest recordings of history will be made. In less than ten thousand years the scientific method will be discovered that made it possible to uncover this tempestuous past.

This dramatic setting of our origins forces a basic rethinking of some of the deepest assumptions about who we are. In particular, it overturns the idea that a single, static environment shaped our brains. Our ancestors lived in a world of intense variability. We owe our existence to how the world's changes and instabilities shaped us as flexible animals. As changing worlds shifted the bare necessities of life, they demanded new ways of living and novel, flexible forms of social organization. Survival of the fittest meant survival of the most flexible.

In order to imagine how these changes would have affected our ancestors, let's ponder for a moment how they might well affect us all in our lifetime. Although global warming has received a tremendous amount of attention, there is an even more ominous threat. If the emissions of a heavily populated and industrialized world continue to perturb the forces shaping our climate, the net effect might not be a warming at all, but a sudden jump back to a cold phase, spurred by changes in the ocean's currents.[23]

The results would be catastrophic. Lacking warm ocean currents, Europe could be thrown into cold within a decade. Since Europe produces food for more people than does North America, the encroaching tundra would cause a devastating crop failure. Economies would topple and regional conflicts would likely erupt as groups fought for resources. Such a large and sudden shift in climate would likely bring as many deaths as a plague.

As we saw in the examples of chimpanzees and bonobos, the game of life changes when climatic changes shuffle resources. The history of life on earth is filled with such stories of climatic betrayal in which species specialized for one climate are decimated by sudden climatic change. During the last million years, climatic change has greatly reduced the diversity of animals in Africa. More versatile species replaced those animals that were too specialized for one way of life, one ecological reality.

We saw that the first step in dismissing the flawed image of our brains as specialized contraptions for an unchanging world was to recognize that our brain's expansion occurred during a period of unprecedented climatic oscillation that was at its most furious during the last six hundred thousand years, exactly when our ancestors' brains underwent their greatest expansion.

How did our ancestors respond to their tempestuous world? A species can respond in a few ways to climatic oscillations. It can attempt to track its habitat as its habitat shrinks and expands with the ebb and flow of climatic change. Frans de Waal suggests that bonobos might have taken shelter in rain forest refugia during glacial periods and thus were buffered from the pressures for change. Another option is to face change head-on by developing a behavioral repertoire sufficiently flexible to prosper in different ecologies. Our ancestors would have needed ways to deal with sweeping ecological changes, sometimes occurring within a single lifetime, at other times spanning many generations. This means that the appropriate way to think about the fitness of a species may sometimes be in terms of its capacity to prosper in many environments. Rick Potts of the Smithsonian Institution's National Museum of Natural History contrasts directional selection, the kinds of pressures that acted on Darwin's finches, with variability selection, the capacity of a lineage to adapt to multiple ecologies. Given the emerging understanding of the earth's climatic history, variability selection looked to be a key piece of the puzzle in understanding the human brain's expansion. It revealed that the human brain's expansion does not signify the accumulation of more and more instinctual behaviors, but rather a growing mental flexibility that expands our behavioral repertoire, a flexibility that lies at the core of who we are.

Coming in from the Rain: Culture as an Umbrella

Variability selection imposes some design problems for evolution. In particular, through what medium can evolution transmit flexible behavior from one generation to the next? Genes are one medium through which a behavior can be transmitted intergenerationally, affecting the design schematics of circuits that generate behavior.

As we considered in the last chapter, however, genes are only one medium of transmission. There are nongenetic mediums as well. Social learning and imitation, for example, are potential ways of teaching behaviors to entire groups, who collectively transmit those behaviors to the next generation. If the goal is to transmit behaviors intergenerationally, then evolution can either build them into the genome or create capacities for their social transmission.[24]

The accumulation of such socially transmitted behaviors is a culture whose change over time is cultural evolution.[25] As Rick Potts suggests, human culture was born as a way to buffer life from a changing and unpredictable world. In Potts's words, culture "is the great fount of human flexibility." Culture and biology aren't two independent mediums of transmitting behavior across generations. Indeed, we have called our own view "cultural biology" to emphasize how both together form a highly dynamic system. They must work together, since biology creates the brain faculties that make social learning, imitation, and other capacities for culture possible. The transmission of behaviors can only work in species that are highly social, suggesting that the ability to acquire a culture is a strong selective pressure. As we will explore in chapter 7, this pressure likely was central in shaping our social brain and creating the extended family.

Thus far, our definition of culture is a minimal one. At a minimum, a culture involves groupwide practices that are passed down from one generation to the next. This means that some nonhuman species are capable of creating culture. For example, some chimpanzee groups can perform complex behaviors, such as nut cracking, that are socially transmitted. This might clash with your intuitions about what constitutes a culture, since many people want to reserve culture to include such elements as language, ritual, and institutions. From an evolutionary perspective, it is reasonable to suppose that the capacities we associate with human culture grew out of more rudimentary ones. Indeed, we would expect that our human capacities emerged out of the long coevolution between biology and culture, each building on and driving the other.[26]

We can look for those rudimentary capacities in nonhuman primates to see more clearly what benefits culture bestows on those who possess it. Recently, researchers from seven field sites where chimpanzees are studied pooled their observations, a combined total of 150 years of observation. They found an astonishing amount of variation in tool technologies and social customs. Some of this variation was due to different ecological conditions, but thirty-nine behavior patterns, from differing ways of using a twig to capture and eat ants to different grooming patterns, weren't simply reflections of diverse ecologies but showed culture's capacity to vary arbitrarily. This indicates that chimpanzees have a remarkable ability to invent new social customs and to transmit them socially. Just as in human societies, chimpanzee groups display rich cultural variations.

The ability to transmit behaviors socially allows chimpanzees to prosper in a wide variety of ecologies, from savanna to dense forests. Tai forest chimpanzees of the Ivory Coast, for example, may spend over two hours a day cracking nuts, a rich source of calories. This behavior is complicated, since it requires cracking open the nutshell without smashing the meat to smithereens. Tai forest chimpanzees accomplish this by using a piece of wood or rock as a hammer and a hard surface as an anvil, a skill that takes many years of practice for a young chimpanzee to acquire.

The capacities that make cultural transmission possible are critical for thinking about how our ancestors likely responded to ecological change, both within an individual's lifetime and across the lineage. Behavioral change in response to shifting ecological contexts is the first fulcrum of human evolution. In the face of ecological instability, our ancestors developed new technologies, expanded their food sources, and learned to use available resources in new ways. We saw that chimpanzees, too, appear capable of inventing new tool technologies, expanding their food sources, and using resources in new ways. These are the signs of flexible and intelligent behavior. At some point, however, our ancestors appeared to have gone in a new direction. Some changes in the brain must have taken place that allowed our ancestors to create more complex forms of social organization, which in turn opened up new cultural possibilities. This leads us back to the prefrontal cortex and its role in understanding oneself and others as persons with beliefs and desires, which in turn appears crucial for the emergence of language and other symbol systems, the signature elements of human culture.

We will explore throughout this book how these capacities appear to be tightly linked to evolutionary alterations to the human brain, allowing us to engage in highly complex social interactions, the second fulcrum of human evolution. These allowed members to build buffers between themselves and the environment by pooling their resources and knowledge for mutual benefit. Ultimately, these capacities allowed our ancestors to create societies that included division of labor, entering into long-term obligations, extending cooperation beyond the bonds of kinship by creating shared, national identities, and accumulating systematic knowledge, expertise, and a historical record.

Complex social organization requires an efficient way to share information, and the most obvious way to share information efficiently is with language. Imagine trying to convey an event in your past, for example, without using language. With language, members of a group can talk about their past, transmit information they learned from their elders, and build knowledge of their world. Language learning appears to depend on a set of social cognitive skills in which we understand others as people with minds.[27] Once children begin to acquire language, it becomes a powerful new tool for elaborating their social understanding, for language mediates our ability to construct models of ourselves and others.[28] (Earlier, we mentioned that autism may stem from an impairment to engage in the social world, thus often leading to delays or impairments in language acquisition, which may further thwart the ability to build social skills.)

Let's return to the story of the young Sebei girl in this context. Why was the young woman at first willing to subject herself to a disfiguring operation? Why was this a public ceremony, and why were members of her group so bent on seeing their traditions continued? The Sebei girl's ritualistic entry into adulthood, like so many rituals that permeate life, defines an abstract social relationship.[29] Most rituals are shared public experiences because they define shared meanings for our social roles. A marriage, for example, is public not only because it involves the public pledge of vows between partners but also because it signals a new set of social obligations among the entire community. It is more than a personal contract between partners; it is a social contract defining expected behavior for all members toward the partners. The Sebei girl's passage into the new symbolic role of an adult member of her community signals her entry into a new symbolic relationship with her group. This new relationship changes how she interacts with others and the range of permissible behaviors.

In our own society, we are constantly troubled over the definitions of our symbolic relationships, and these relationships are increasingly legalistic. When, for example, should a young offender be considered an adult? Why do we care? We care because an adult is more than a physical definition, it is a symbolic role, carrying with it certain expectations. We expect adults to understand the consequences of their actions and to be able to restrain their impulses. When they transgress these expectations, we hold them responsible. Each of us, whether we're aware of it or not, plays multiple symbolic roles, whether as parent, spouse, grandparent, or professional. Each role defines how we interact with one another and our range of permissible behaviors.

Constructing Yourself: Some Self-Assembly Required

The models of self and others that we acquire through our immersion in language and culture can be thought of as a user's guide to life: They allow us to respond flexibly to different contexts, and they teach us how to regulate our own behavior. We've seen that extending childhood is a way to acquire a broad and flexible repertoire of behaviors, thereby escaping the limits of fixed instincts. But something further happens in humans—likely mediated by such regions as area 10 of the prefrontal cortex—that makes our cultural worlds possible. As a child interacts with the world, she is building richer and richer theories about the world. Soon, she will learn that these experiences belong to a single person. She will gain the ability to detach from her immediate context and envision life extending both in the past and in the future. She will have created an enduring self. It is her life story, her identity. Eventually, she will even recognize the inevitability of her story's end, and such thinking will produce the human emotion of anxiety. Her symbolic self will also create the uniquely human emotion called shame that may propel her—as happened with the Sebei's girl bid for suicide—to prefer feeling nothing at all.

Driven by your interactions with others and the worlds of meaning you share, you literally construct yourself during childhood. It is no accident that your brain's development is so protracted and that the last parts of your brain to develop are those underlying self-assembly. Your user's guide to life isn't wired into your genes; it is human-made, with major elements of it held in your culture. It takes a lot of experience and practice to construct these models and to learn the rules of the game in your culture. Your long childhood, making you so vulnerable and requiring so much parental investment, gives you extraordinary flexibility.

The idea that human development is a process of enculturation sounds as though the developing child passively absorbs a culture. The crux of cultural biology is that our biology makes us an active partici-pant in this process. As we will explore in the next chapter, your biol-ogy has primed you to acquire a culture. It has endowed you with an "internal guidance system" that propels you from within and bootstraps you into culture by making the social world highly significant and fuel-ing your desire to participate in it, a desire that was so strongly illus-trated in the case of the Sebei girl.

Culture's Faustian Bargain

The capacity to construct models of self and others, then, is the bridge that carried our ancestors over the mental Rubicon. It is the founda-tion for human culture—language, complex social behavior, ways of conceiving the world, and institutions—that is key to who we are.

Although the brain-culture interaction is a great source of human flexibility, in later chapters we'll examine how it also creates an almost pathological need for meaning and coherence. This can be a powerful source of creative energy, driving both science and the arts to ponder our world and our place in it. But we also pay a high price for our abil-ity to create ourselves. The prefrontal cortex that underlies many of our symbolic abilities is also involved in virtually all mental illnesses, in-cluding schizophrenia, depression, obsessive compulsive disorders, and manic states. This chapter opened with the story of a young girl seeking solace in suicide. As much as we worry about murder rates, we forget that more people die by their own hands than by the hands of others. This, too, is uniquely human, with roots in our prefrontal cortex.

Our protracted developmental dependence also makes us suscepti-ble to disruptions in the journey to becoming a person. Under proper guidance, our natural inclinations blossom into our capacity for empathy, our ability to take someone else's perspective. When someone's journey becomes derailed, however, it is difficult for the traveler to make sense of and learn from the experience and to see it from another's perspective. All of the attention paid in recent years to human acquisition of emotional intelligence boils down to how well we learn the social rules of our culture. We have a name for people whose impairments of self manifest themselves in an inability to feel empathy: sociopath. We will explore the disturbed inner world of these people in chapter 8.

The Well-Cultured Brain

Despite the vulnerabilities of a human mind, its advantages are obvious. A tempestuous history of jarring climatic instability has caused our brains to thrive on change itself. How does your brain use culture to build a user's guide to life, or what in everyday conversation is called your personality or your self? The next step of our journey will take us into a strange land, to the nethermost regions of the brain, where we will discover our internal guidance system, that deeply buried core where thought and feeling intertwine to build our sense of self.

Between Thought and Feeling 5

The Mystery of Emotions

t's late and you're channel surfing. You flick, or we should say electronically forage, to an old *Star Trek* movie. Mr. Spock is back on his home planet spending his days in quiet meditation, after having been recently rescued from death by his friends. In fact, they went one step further and brought him back to life—such is friendship in the twenty-third century. Yet Spock is strangely indifferent. His earlier dry wit all but gone, he can't understand why his friends acted so illogically. He just mumbles about the needs of the many outweighing those of the few. Without emotions, he seems incapable of friendship or being a part of the group. He has become a loner, a hermit.

Deciding that watching the inner anguish of a Vulcan is like watching a lobotomized Woody Allen, you embark on some more electronic foraging. The Home Shopping Network comes on and there's the motivational speaker Anthony Robbins selling his latest video series. But what strikes you is how different he is from Mr. Spock. Robbins is gushing with enthusiasm. His movements, gestures, and expressions are an exaggerated orgy of excitement. He's making you feel good just watching him be so happy. Although Robbins is selling motivation, we'll discover that his happy face is no accident. If Mr. Spock is the icon of remote intellect, then Robbins is the icon of gushy emotion. But why do you have emotions at all? Spock would point to the human foibles that stem from your emotions.

A cornerstone of thinking about who we are is the idea that we can cleanly divide thought and feeling, emotion and intellect. Mr. Spock is, in fact, just a modern representative of an archetype that goes back at least as far as Plato, who likened the human mind to a chariot drawn by two horses—one intellect, the other emotion—together an unruly team. While intellect drove up toward the rarefied air of reason, emotion drew the chariot down to the base world of the sensual. Its driver was in a never-ending struggle with the pair, always fighting to rein them in. Nineteenth-century Romantics revived a pair of Greek gods, brothers Apollo and Dionysus, to represent the warring duo: Apollo, calm and serene; Dionysus, wild with drunken excess.

Beyond being illogical, as Spock would charge, our emotions come at a real cost. Virtually every mental disorder is accompanied by emotional disturbances.[1] And of the ten most expensive medical conditions in the United States, mental disorders rank third—about $150 billion is spent annually on health care for mental disorders (not including expenditures for drug abuse treatment).

Mood disorders, known clinically as affective disorders, are among the most widespread mental health problems facing us. As much as 20 percent of the American population will become clinically depressed during their lifetime. Depression strikes women more than men, perhaps because of certain differences in their brains.[2] Beyond the direct health risks, there is now increasing evidence that depression heightens the risk of death after heart attack, stroke, and cancer. About two million Americans suffer from manic depression, or what is clinically called bipolar disorder, characterized by severe mood swings that can last from one week to many months. Partly because of the enduring social taboo surrounding mental disorders, about a third of those with bipolar disorder will go untreated. Of these, nearly a fifth will end their life by suicide. Suicide itself is the tenth leading cause of death, and its rising trend among the young is especially disturbing. There must be some good reason for having emotions that outweigh these grim statistics. But what is it?

As we were working out our theory of constructive learning, we began to see that the brain's emotional systems play a crucial role in building the symbolic mind during development, and indeed play a lifelong role in guiding our decisions, suggesting that the mind is organized very differently from how some cognitive psychologists see it—a difference that holds far-reaching implications for how we understand ourselves. In this chapter we look inside the brain to search for where your emotions reside. There, we'll discover a mysterious tangle of brain systems that generate your emotions, moods, and motivational drives. As we shall see, getting a glimpse into our essence means asking why a child laughs four hundred times a day, an adult twenty-five, and Spock zero.

Emotion's Ancient Roots

What is so surprising about this tangled web is that its roots are in ancient brain structures whose basic functions can be found in animals as diverse as bees and humans. We took the fact that these structures are so highly conserved across species to suggest that they must hold some essential truth about how brains work. For example, one structure, called the ventral tegmental area (VTA),[3] is buried in the deep recesses of your brain in a region called the basal ganglia, which lies beneath the cerebral cortex, and is part of a powerful reward system; indeed, this is the system that drugs of addiction hijack. We postulated that perhaps emotions, motivation, and reward are all facets of a single, powerful way to design brains.

Computer models of the VTA in Terry's lab were revealing that this brain area could learn to predict future reward based on past experience. For example, if you perform well at work, you might expect a bonus at Christmas, as happened last Christmas. We realized that the VTA and related structures are your internal compass. It fills your world with values, provides emotional tone to your experiences, and helps you decide what fork in the road to take when you face decisions. It is your internal guidance system, creating desires, propelling you to action, and helping you get on in the world by predicting the benefits of possible decisions. It is why you are not just an inert piece of clay to be molded by the world, or a collection of genetic reflexes optimizing fitness.

The intrigue deepened when we discovered that the VTA sends its fibers to the prefrontal cortex, which is a key to your humanity, mental flexibility, and sense of self. How and why was an ancient structure, the VTA, intertwined with the prefrontal cortex? An important clue came from development. Damage to the VTA impairs the development of the prefrontal cortex and leads to severe mental retardation.[4] We realized that this ancient structure is at the core of a tangled web that bootstraps us into culture and helps construct the human mind.

The VTA first sparks to life in the child. It is the vital force that propels you to actively engage the world with the human texture that a Spock could never know. Evolution shaped your brain not by departing from the powerful reward-based systems that you share with the bees, but by adding new layers onto this basic architecture and extending its development.[5] Even in adulthood, these two intertwined brain systems operate in close concert. Together, they underpin your everyday thinking, planning, and decision making. Indeed, their joint workings suggest that the age-old dichotomy between thought and feeling, intellect and passion, is a faulty one, carrying important implications for your inner life.

This, then, is the basic view of the mind that cultural biology re-

veals, a sort of *Upstairs, Downstairs* in which old and new come together to create your immensely complex behavior.[6] The human dramas that we will encounter in later chapters have their roots in the
interplay between the VTA, other subcortical structures that utilize the
chemical messenger serotonin, and the prefrontal cortex, and in the
tension between old and new components of this normally integrated
behavioral system. It is at once ancient and new, a far more powerful
way to generate flexible behavior than genetic reflexes and far stranger
and more fascinating than Plato or psychoanalysts ever imagined.

The Serotonin Connection: From Here to Ecstasy

Matthew Klam was an unhappy and angry college student. Soon after
taking Ecstasy for the first time, his clothes began to feel softer, music
sounded better, and soon he was feeling much better about himself
and his friends.[7] Also called MDMA,[8] Ecstasy has become the drug of
choice at all-night raves, where crowds of dancers lose themselves in a
cloud of sociableness.

Ecstasy causes a massive release of serotonin in the brain. Serotonin is one of the brain's major chemical messengers; it influences
mood, appetite, sleep, and other important functions such as heart
rate, blood pressure, and body temperature. The brain of every vertebrate, including ours, has cells that manufacture serotonin using the
amino acid tryptophan as a raw material. These neurons are located in
a part of the brain stem known as the dorsal Raphe nucleus, a small
cluster of neurons. This is an ancient part of the vertebrate brain that
regulates arousal and attention. What is so intriguing about this center
is that it sends its outputs widely across the brain, influencing billions
of neurons in the cerebral cortex, the most highly evolved part of the
human brain.[9]

After the effects of Ecstasy wear off, some four to six hours after
taking it, nonsociability, fatigue, and irritation replace the sense of happiness Ecstasy first brings on. Ecstasy releases most of the available

serotonin in the brain and it takes weeks to fully replenish. If taken more often than once a month, Ecstasy begins to have a much smaller impact, likely because it is exhausting the serotonin system.

Recreational use of Ecstasy is on the rise across almost every ethnic and class group. Most Ecstasy is made in clandestine labs in Europe and smuggled into the United States. Seizures of Ecstasy by the Customs Service have jumped sharply, from only 400,000 pills in 1997 to 9.3 million pills in 2000, worth $200 million on the street. Short-term side effects of Ecstasy include increased blood pressure, heart rates, and body temperature, resulting in dehydration and hypothermia. The use of Ecstasy also leads to a lowered number of serotonin receptors, a process of downregulation that takes weeks to occur and lasts many months. The prolonged administration of Ecstasy causes serotonin nerve terminals to degenerate in animal studies. The long-term effects of Ecstasy on humans are not known, but those who are taking it recreationally may soon discover the answer, an experiment that would have never been approved by human subjects committees.

Prozac, a drug that is effective in treating some types of depression, also increases serotonin activity in humans.[10] More than seventeen million Americans have taken Prozac, which for some has been a lifesaver.[11] Even many who do not suffer from clinical depression find that it lifts their mood, making them feel more self-confident and more sociable. In contrast to Ecstasy, however, those who take Prozac only begin to feel better about themselves after many weeks of taking it, and the effects last much longer. The key difference between Prozac and Ecstasy appears to be that Ecstasy alters serotonin levels rapidly and depletes its stores, whereas Prozac causes slower and longer-term increases in serotonin activity. The long delay in the impact of Prozac is not yet understood, but it may involve the genetic regulation of serotonin signaling between neurons.

In contrast with the positive effects of increasing serotonin activity, reduced serotonin activity has negative effects that include depression, suicide, risk-taking behaviors, and impulsive violent behaviors.[12] In mice that have been genetically engineered to lack one type of serotonin receptor, the mutant male will immediately attack an intruding male; in comparison, unmodified males first go through a period of observation, sniffing, and aggressive displays such as tail rattling.[13] The number and intensity of the attacks was also much higher in mutant males. Similar types of impulsive aggression have been observed in monkeys with low serotonin activity.[14] There is even evidence that cholesterol-lowering drugs, now the largest-selling class of drugs in the world, may lower serotonin activity as one of their side effects, and in so doing increase the rate of accidents, suicide, and homicides—violent deaths.[15]

Today, the idea that brain chemistry underlies mood is taken for granted. However, it also raises troubling questions about who we are, in part because people who took Prozac didn't feel that the drug just lifted their blue mood, but that it often transformed their personality. Many people started having feelings of greater self-worth and confidence. They became less sensitive to criticism and social rejection, and more willing to stand up for themselves. Did Prozac unmask people's "real" self, which had been obscured by a chemical imbalance? Or was it the first step toward a future filled with "designer personalities"?

We have seen only the first few hints of the ways personality can be altered with drugs. Using emerging genetic techniques, in the future it may even be possible to instruct brain cells to make new kinds of chemical messengers and receptors. Imagine designing supermotivated, elated personalities, or complacent "worker bee" personalities. The inextinguishable human spirit that topples an oppressive order in *Brave New World* is, after all, ultimately the result of a chemical soup inside the brain, and there's nothing to guarantee that the recipe can't be changed.

But let's leave aside these future possibilities and explore the present. As we noted, Prozac primarily alters levels of the brain molecule serotonin. So questions about Prozac lead to questions about what serotonin does, and looking at serotonin gives us our first glimpse inside the tangled web of the human core.

Consider the laundry list of maladies serotonin has been implicated in: uncontrollable appetite, obsessive-compulsive disorder, autism, anxiety and panic, premenstrual syndrome, migraines, bulimia, social phobias, lack of motivation, low social status, unsociability, schizophrenia, and violent rage.

Abnormal serotonin activity also appears to underlie bizarre rituals, such as constant hand washing or checking to see if the front door is locked, and to interrupt some people's attempts at normal life. Deficits in serotonin levels in a part of the brain known as the striatum might be involved in obsessive-compulsive disorder. The striatum organizes sequential movements, such as walking, whose repetitive motions stimulate the release of serotonin. The repetitive motions seen in obsessive-compulsive disorder might be a form of self-medication to boost serotonin levels.[16] It might also underlie some of our nervous habits and even gum chewing, raising the possibility that those who can walk and chew gum at the same time might get a double dose of serenity.

Serotonin is involved in other forms of altered states as well. Along with Ecstasy, drugs such as LSD and mescaline also activate the serotonin system in the brain. For those who went on mind expansion quests, from Timothy Leary to Carlos Castaneda, it looks as if their slogan should really have been "turn on and tune in" to serotonin.

Serotonin, then, is involved in many essential facets of being human. What we found so surprising, however, was that the serotonin system, underlying essential human qualities, is among the oldest of brain systems; it is found in the most ancient invertebrates and in every vertebrate brain. It seems that nature got it right and hasn't changed it much since. Your brain contains a meager hundred thousand or so cells that make serotonin. While that may seem like a lot, it amounts to around 0.0001 percent of the brain's total neurons. Found in your brain stem, serotonin-producing neurons send their axons sweeping out over your entire brain to influence virtually every neuron.

One popular way to think of the release of serotonin at nerve terminals is as garden hoses that have small holes punched all along their length. These "hoses" sprinkle your brain with serotonin, which doesn't so much give cells their marching orders as change what they're already doing. Envision a single molecule of serotonin as a ball in a pinball machine. Once released, a serotonin molecule bounces around until it finds a receptor it fits into, akin to the hole a pinball temporarily falls into. Once bound to a receptor, serotonin sets off a cascade of changes inside the cell, including how it reacts to other cells trying to communicate with it. After its brief stay inside the hole, serotonin is spit out to do some more bouncing around. And the bells and whistles going off inside your head in response are no less dramatic than a pinball machine's flurry of activity.

However, a pinball can't be kept in play forever. Eventually, the gaping hole at the bottom of the machine swallows up the ball. Inside your brain this gaping hole takes the form of reuptake mechanisms on the cell that released the serotonin molecule. This cell traps the serotonin and sucks it back inside, where it is repackaged and put back into the queue for another play. The reuptake mechanism is where Prozac has its effects. You can think of Prozac as a blocker that sits over the gaping hole of a pinball machine, keeping the pinball in play longer; thus, its technical moniker as a selective serotonin reuptake inhibitor (SSRI).

The serotonin system isn't the only mood system in the brain stem that fans out over wide areas. Another uses a chemical called noradrenaline (also known as norepinephrine). A drug, reboxetine, marketed under the name Edronax, works along the lines of Prozac, except that it targets cells that make noradrenaline. Reportedly, it is more effective than Prozac in helping patients suffering from anergic depression, who lack capacity for sustained motivation, return to productive lives. Whereas serotonin activation results in a serene mellowness, noradrenaline activity is more involved in increasing drive and motivation. When a closely related molecule, adrenaline, is released into your blood, it makes your heart pound faster and prepares your body for vigorous physical activity, a counterpoint to the increased mental activity triggered by noradrenaline inside the brain.

It may seem paradoxical that targeting more than one brain system can treat depression. Do imbalances in serotonin or noradrenaline cause depression? We now know that these ancient structures interact with one another like players in an orchestra, so part of a serotonergic drug's efficacy may be indirect by altering levels of noradrenaline. This interaction complicates the search for effective treatments for mental disorders and reminds us that, although we talk about this or that chemical system in the brain, we can't lose sight of the fact that brain systems are deeply integrated.

The realization that noradrenaline is linked to motivation took us to the next step in uncovering the mystery of the tangled web and why Tony Robbins is so different from Mr. Spock.

What Makes Spock Run?

Not even Bones, the feisty ship's doctor in *Star Trek,* was able to figure out Spock. From the vantage of human psychiatry, the real problem with Spock is why he does anything at all. Had Bones been a twentieth-century psychiatrist, he likely would have diagnosed Spock as clinically depressed, perhaps suffering from anhedonia—nothing seems to give Spock pleasure. For him, life has just lost its fun. No doubt a twentieth-century psychiatrist would have put Spock on Prozac.

What gets Spock out of bed in the morning? No matter how impressive Spock's thinking may be, there doesn't seem to be anything inside him to get him into gear. Is there a connection between Spock's lack of emotion and his lack of motivation?

We can only speculate about the inner world of a Vulcan, but the question is easier to answer for humans. Consider the faces of bipolar disorder.[17] People who are bipolar often feel their mood is elevated while in the throes of a manic phase, usually to the point of a nervous irritability. They typically talk quickly and excitedly, as though they are struggling to keep up with the thoughts racing through their head, which often make it hard for them to concentrate. Most worrisome is the obsession with pleasure seeking, often with frighteningly little concern for the consequences. In the midst of a manic phase, many work on wildly impractical business schemes. Their shopping sprees rack up huge credit card debts. When in the trough of depression, they have little energy and few goals. They lose interest in the things that used to be so enjoyable. They often lose their sex drive and instead find their thoughts revolving around gloomy images of illness and death.

As the sufferers of bipolar disorder reveal, emotion and motivation are tightly intertwined in us. More than twenty-five years ago, Jeffrey Gray helped revolutionize the study of emotions when he suggested that emotions stem from our brain's motivational systems.[18] Motivation is basic to survival and the pursuit of life's goals. This gets us closer to understanding why Anthony Robbins is in such a good mood on TV. If we asked you to describe Robbins, chances are you'd pick words like gregarious, talkative, enthusiastic, cheerful, peppy, bold, lively, and optimistic. These are all words that describe a personality trait: extroversion. Remember, Robbins is a motivational speaker. It is not a coincidence that most of the words describing his positive emotions also describe positive motivation. If we were to scan Robbins's brain in search of the source of his positive energy, one system that would undoubtedly light up is what's known as the "behavioral facilitation system."

This motivational system, which uses the chemical dopamine, is centered in a small group of cells in the midbrain, in the VTA mentioned earlier and the *substantia nigra pars compacta*.[19] When we saw how this system of dopamine neurons underlies both the motivational and the emotional facets of extroversion, facilitating our pleasurable engagement with the world, we realized that it plays a central role in cultural biology.

Cheer Up, Things Could Be Worse

Dopamine is involved in our everyday good moods, what researchers clinically refer to as positive emotionality. The grim mental health statistics we cited earlier might have given you the impression of a gloomy nation. When people are asked to guess what percentage of Americans are happy, the typical answer is somewhere around 50 percent. The real figure, however, is a little over 80 percent.[20]

There's no question that our triumphs and tragedies color our emotional life.[21] But there seems to be a set point to our emotional thermostat that's adjusted to a positive value in most of us. For most people, it's set at a comfortable seventy degrees Fahrenheit. The slings and arrows of life, along with good fortune, are like winter storms or summer heat waves that cause fluctuations, but the temperature typically bounces back to around seventy degrees Fahrenheit, or whatever your comfort zone happens to be. Surprisingly, the vast majority of people facing hardships ranging from unemployment to severe physical handicaps report that they are happy, despite objective physiological evidence for high levels of stress hormones in their blood. And while our media tend to concentrate on human fragility, people are remarkably resilient, able to endure the horrors of war, concentration camps, and other terrible ordeals to live normal lives.[22]

On the other hand, lottery winners, who according to the ads go on to live happily ever after, do report a brief euphoric period, but usually within a year they say that they're no happier than before they won. Hence, even the rewards of good fortune ultimately diminish, as we return to our emotional set point. Above a certain level, the happiness of a nation does not rise with its wealth and may actually decline.[23]

Most people guess that at least half the population will become clinically depressed sometime in their life, but the actual figure is about 10 percent. That's 5 to 12 percent for men and 10 to 25 percent for women. We'll take a longer look at happiness in chapter 10. For now, let's just say that for most of us it really does seem it's good just to be alive.

Why do many people—despite often huge obstacles—have their emotional thermostat hovering around a positive setting? The clue is in remembering that the same system underlying our emotional thermostat similarly underlies our drive and motivation, also called our approach tendencies. Here's what psychologists Ed and Carol Diener say:

[B]ecause positive moods energize approach tendencies, it is desirable that people on average be in a positive mood. Human approach tendencies are manifest in the rapid exploration and settlement of new frontiers and in the unremitting invention of new ideas and institutions throughout human history. Thus, not only might humans' large brains and opposable thumbs be responsible for the rapid spread of humanity across the globe, but positive emotions might also be an important factor.[24]

In other words, being in a good mood energizes some people to seek out new worlds and to boldly go where no one has gone before. Poets throughout the ages have rhapsodized about the fires burning that animate these actions. We now know that brain messengers like serotonin, noradrenaline, and dopamine fuel these inner fires. Indeed, these ancient structures underlie major dimensions of human personality and our mental health, together orchestrating the richly textured score of the human core.

If, however, we share these ancient systems with other animals, such as mice and bees, how could there be anything uniquely human about them? To see what makes up the human core, we'll need to examine how these ancient structures fit with other parts of the human brain to generate human qualities. Indeed, the fact that these structures have been a part of brains for so long indicated to us that they must reveal something very basic about the design of behavioral systems and why we have emotions at all.

Designs for a Brain

As we've mentioned, being in a positive mood and being motivated to pursue goals are rooted in a common brain system, the dopamine system.[25] That's why Tony Robbins smiles so much. But the more basic question is, Why are we interested in interacting with the world at all? In other words, why do we pursue goals and why do certain facets of the world have deep rewarding value to us?

Consider the life of the sea squirt. As a newborn, this little ocean dweller swims off in search of a good place to live, all the while struggling to feed itself. When at last it finds a habitat, a little nook or cranny in a piece of coral, it backs in and permanently attaches itself. No longer in need of its brain to get on in the world, it eats it. As the philosopher Dan Dennett remarks, it's a lot like getting tenure.

The humble sea squirt illustrates the most basic reason why we have a brain. Creatures need a brain to move around a changing world to obtain life-enhancing goals. Two features of our changing world make brains possible: Change is uncertain but predictable, and uncertainty goes primarily in one direction—the future. If change happened randomly, then predictions about the future would only be as good as chance—hardly a basis for life. Fortunately for all creatures, the world changes in uncertain but predictable ways. It's also fortunate that a truth about the matter is established once things happen. This allows creatures to build knowledge of the past.

These turn out to be the essentials of brain design: a database of past events and a mechanism to use that knowledge to make predictions about the future to guide action to obtain life-enhancing goals. Brains are prediction machines that use information gathered from past experience to predict future events important for survival. This is one of the few things that all brains do, from sea squirts' to honeybees', humans', and even the autonomous robots' that will likely take part in space exploration one day.

The incredibly complex human transactions we witness every day in our modern society reflect this truth. Suppose, for example, that you're traveling in a town you've never been in before and you see a Starbucks sign on the corner. You'd be pretty certain what to expect there because you've learned what the sign signifies. But if you lived in a town where the shop signs are rearranged every night, you'd have no way of finding your morning cappuccino, to say nothing of guiding your life. So we owe our brains to a world of regularities that make predictions possible.

The Nobel Prize–winning philosopher Albert Camus once wrote that to breathe is to judge. Whether or not we are judging life worth living every moment, your brain is making life-preserving decisions every moment. Most occur without your noticing them. But like Hamlet's quandary over how to avenge his father's murder, the heart of every decision lies in making predictions about the future. For Hamlet, that meant imagining various "what-ifs," which unfortunately he was better at doing than making decisions.

The rich, textured bundle of memories and feelings that are you are there for one basic reason: to guide you as you face an uncertain world. They don't exist just so you can reflect on your past, like Swann does in Marcel Proust's *Remembrance of Things Past,* although that's part of the story. We are obsessed with finding patterns and making sense of what happens around us because we need that information to predict the future well enough to make good decisions. It's probably no accident that many schizophrenics construct grand plots because their minds are working overtime, creating stories that have become detached from reality. But even the rest of us aren't immune to doing this. Many people are vulnerable to believing conspiracy theories, whether it's the second gunman theory in JFK's assassination or aliens at Roswell. We all work under a belief that the truth is out there.

This brain design is very different from one based on building in specialized circuits that generate specific behaviors, otherwise known as instinctual behavior, to obtain life-enhancing goals. A prediction machine model of brain design leaves the way to life-enhancing goals open and gives the system ways to learn the best path to achieving them.

This is an extremely powerful way to build intelligent life. How could a brain learn how to behave in a life-enhancing way? The key, we saw, is to infuse life-enhancing goals with rewarding value and to endow a brain with ways to predict the reward value of its actions. Once it can do this, then it is a matter of choosing the action with the highest return. This also provides the basis and incentive for learning. Once a brain can predict the amount of reward it will receive, it can calculate the difference between its prediction and the actual reward it receives when it performs the action. This is why the brain systems of prediction and reward are so intimately related. As we explore in the next section, they form the basis of an extremely powerful form of re-inforcement learning and are a key to unraveling the human core and our emotional life.

Deciding to Survive

For many species, survival depends on getting the next meal. If you were a worker bee, you would spend your days preoccupied with what flowers to visit for drops of nectar. A single trip may take the bee sev-eral miles, and during its brief lifetime a worker bee will travel hun-dreds of miles. Bees initially have a preference for blue flowers, which to an evolutionary psychologist might suggest that bees have a special nectar module for blue flowers. But bees, which are among the smartest insects, learn to associate a flower's odor, shape, and color with nectar. The more nectar, the greater the reward, and the more likely the bee will return to that patch of flowers. Bees use these pat-terns of reward to guide their foraging.

In 1995, Terry visited Martin Hammer at the Free University in Berlin[26] and was impressed by Martin's discovery that a single neuron in the bee brain underlies the association for odors. This cell, called VUMmx1, uses a chemical called octopamine, which is chemically similar to dopamine, and is part of a decision-making system whose function is similar to that of the dopamine system in vertebrates.

In Terry's lab at the Salk Institute, Read Montague and Peter Dayan realized that examining the octopamine system in the bee might shed light on how similar systems work in us. In particular, they wanted to understand how this system guides learning. Pavlov and others, who demonstrated that animals could learn new patterns of reward by conditioning, had long ago discovered the link between re-ward and learning. Yet most contemporary psychologists dismissed re-ward learning as an impoverished form of learning. Researchers in computer learning, however, had shown that new forms of reward learning were promising, so Montague and Dayan wanted to know whether biological systems utilize complex forms of reward learning.

One intriguing possibility was that animals use a reward signal as a teacher. Consider the following analogy. Almost everyone has played the child's game of hotter/colder. One child begins to take small steps trying to find a hidden reward. Another child calls out "hotter" when the first child steps closer to the target and "colder" when that child takes the wrong step. It's a very simple way to navigate, but it's remarkably effective.

To see whether something like this goes on in bees, Montague and Dayan made a computer model of the bee's octopamine reward system based on how this system works in real bees. When a bee is learning what, in its world, are "hotter" and "colder" odors for reaching its goals, it is turning remembered rewards into predictions about the prospects of future rewards associated with different odors. The simulated bee behaved remarkably like real bees, capturing a wide variety of their behavior, including the risk-aversive behavior bees often display in avoiding patches of flowers where the nectar fluctuates highly from flower to flower. The success of these computer simulations led Montague and Dayan to suggest that the output of the VUMmx1 measures prediction error.[27] Its output rises when the bee gets an unexpectedly bigger drop of nectar from a flower, in essence saying to the rest of the bee's brain: You're doing better than expected, so remember that the color, odor, and shape of this flower predicted this amount of reward. In this way, the bee uses its experiences to store predictions of the future to guide its behavior, like steering toward a particular patch of flowers. It is teaching itself about its world.

This strategy is similar to a technique used in building artificial learning systems known as Temporal Difference (TD) learning, a type of reinforcement learning developed by computer scientists Richard Sutton and Andrew Barto.[28] Although learning often connotes the presence of a teacher who tells us whether our work is correct, most learning involves a subtler teacher: the feedback from the consequences of our actions. TD learning uses just this kind of feedback by using the difference between predicted rewards and actual rewards to guide adaptive behavior.

The fact that bees have a system that uses exactly this prediction signal suggests that nature probably invented TD learning long before it was invented for computer systems. What role could similar ancient systems have in our mental life? Could your behavior be driven by the same type of predictive learning found in the bee?[29]

Games Computers Play

You may think that this apparently simple form of learning seems too simple to lead to interesting or complex behaviors, such as the ability that many animals have to navigate in a complex environment or plan a sequence of steps to achieve a goal.[30] Your intuition, though, may change after you consider a striking demonstration of the power of this type of learning and the important lesson it holds for thinking about yourself.

Let's first consider the famous chess machine, Deep Blue, which made international headlines in 1997 when it beat Garry Kasparov. Researchers in artificial intelligence have long tried to make machines that beat humans in chess and other games of skill. Beating a world chess grandmaster had been a holy grail of artificial intelligence since the 1960s, and after a long string of defeats, IBM decided to use a brute force approach to crack the problem. Built around computer chips that were specialized for making chess moves and an array of thirty-two powerful computer processors, Deep Blue could analyze thirty-six billion moves in three minutes, something that would take Kasparov almost four hundred years of around-the-clock thinking to do. This meant it could search much deeper into the number of all possible moves, which are essentially unlimited.[31] All this computing power unnerved Kasparov during their first tournament because it gave the appearance that Deep Blue was pursuing a game strategy the way good human players do.

For all of its enormous power, however, Deep Blue is as preprogrammed as an actor following a studio script. For Deep Blue, a team of computer programmers and world-class grandmasters wrote the script, its knowledge of chess emerging full-blown as if through an innate evolutionary module. This is the antithesis of learning through experience. Given Deep Blue's computing power, it should have trounced Kasparov, but in fact he lost only 2.5 to 3.5 games to Deep Blue, indicating that humans don't play chess the way Deep Blue does.

While Kasparov was battling Deep Blue, a quieter and even deeper battle was occurring at IBM that may have far-reaching consequences for how we may interact with computers in the future. This battle wasn't between computer and human, but was waged entirely within a computer. TD-Gammon, a program written by Gerald Tesauro at IBM's T. J. Watson Research Center, was teaching itself how to play backgammon. Unlike chess, backgammon—like life—is a game of chance, since players roll a pair of dice at each turn to determine how far they can move their pieces in a race to the finish line.

In 1987, Tesauro, then at the Institute for Advanced Study in Princeton, and Terry, then at Johns Hopkins University, trained a neural network to play backgammon based on an expert's assessment of the best move from thousands of board positions.[32] The network, called Neurogammon, played at an intermediate level, which was surprising, since there are over 10^{20} possible board positions and Neurogammon learned from only a tiny fraction of these. Somehow, however, it was able to transfer what it had learned to new board positions and made the best move most of the time. There was a problem with this way of teaching Neurogammon, though: The program could never play better than the level of its teacher.

After moving to IBM, Tesauro went on to develop a new approach to computer backgammon that would go beyond the limits of Neurogammon. He blended two techniques that had been developed earlier. To eliminate the teacher, he turned to self-play, a technique that A. Samuels developed at IBM in the 1950s for computer checkers. This way the program could play itself over and over, leaving no limit to how good it could get. The second technique involved finding a way for the program to improve when the only feedback it received was the reward or punishment—win or lose—that came at the end of each game. The problem, known technically as "temporal credit assignment," is one we all face when playing backgammon. After we've won or lost, we often wonder where we went right or wrong, what our good moves and what the bad ones were. A computer playing itself must be able to evaluate its own moves if it is to improve.

Tesauro gave the program the same learning scheme that Montague and Dayan used in their model of bee foraging: TD learning. Equipped with this form of learning, TD-Gammon started from scratch and played randomly because it had no built-in knowledge of the game. TD-Gammon worked by predicting the game outcome prior to making a move. Once it made the move, it then made a new prediction of the game outcome. This allowed it to measure the difference between its prediction before and after a move to determine how well its prediction before moving really was. You've probably had the experience of predicting that a game move would bring about victory only to be surprised—and dismayed—by your opponent's response. That difference between your prediction of a move's value and its actual value is a powerful teaching signal, and it is what allowed TD-Gammon to learn to play. For TD-Gammon, as the predictions became more and more accurate with experience, information about winning or losing at the end of the game propagated toward the earlier part of the game, allowing it to greatly improve. After a few thousand games against itself, it began to play at a beginner level, literally pulling itself up by its bootstraps. After a hundred thousand games, TD-Gammon could beat Neurogammon. After a million games, it could play at a master level, on a par with the best human players in the world.[33]

Something interesting has happened in the world of backgammon. Human experts are sometimes surprised by the choices TD-Gammon makes. But after extensive analysis, they typically find that TD-Gammon made the right move—that it had discovered, on its own, entirely new strategies. Unlike Deep Blue, TD-Gammon has a learning intelligence that humans not only appreciate but can learn from. When this kind of flexible learning is put into the machines that surround us, we can expect some real intelligence from them.

TD-Gammon and the model of bee foraging provide a crucial insight into how intelligent systems can be built. Although your intuition might be that complex behavior can only be the result of building in highly specialized knowledge, as in Deep Blue, it is possible to build more flexible and sophisticated systems by letting them interact with a rich environment and get feedback from it. In the case of TD-Gammon, even a weak-looking signal—a simple win or lose at the end of the game—can produce intelligent behavior over time. One of the important lessons from TD-Gammon is that it works as well as it does.[34]

How You Learn the Game

So far, we've looked at how the octopamine system works in bees and how engineering principles based on that system could be used to produce complex systems like TD-Gammon. Since the dopamine system of mammalian brains is closely related to the octopamine system in the bee, our next step was to examine whether the principles of bee brain design apply to humans.

Contemporary theorizing about the dopamine system goes back to the 1960s, when psychologists implanted electrodes into the dopamine system of rats.[35] The rats were given access to a bar that would deliver a pulse of electrical stimulation to their dopaminergic neurons whenever they pushed it, thereby releasing dopamine and activating neurons that respond to dopamine. The rats wanted to do little else than self-stimulate their dopamine system, even preferring it to eating and sex. So strong was this desire in the rats that researchers felt they had located the brain's pleasure center.

As the dopamine system was investigated more thoroughly, researchers found that it is involved in all forms of drug addiction, including alcohol, heroin, amphetamines, and nicotine. In 1994 alone, 165,000 deaths were blamed on tobacco-related cancer, highlighting the enormous power nicotine holds over people despite grim warnings of tobacco's consequences. This is also the system where cocaine has its effects, which are powerful enough to hook at least two million Americans. Cocaine interferes with a vacuumlike molecule that sucks up dopamine once a cell has released it, leaving dopamine free to activate the cell over and over.[36] Rats hooked up to a pump that delivers drugs like cocaine, heroin, and amphetamines forgo willing sexual partners for the chance to self-administer these drugs. Indeed, rats, like some people, neglect the normal necessities of life for a fix even to the point of death.

Accumulating facts that refuse to fit into existing theory typically sparks change in science. So it was with the dopamine pleasure principle. Starting around 1990, researchers began wondering aloud about some paradoxes of the pleasure principle. One of these concerned the behavior of schizophrenics. If the amount of dopamine in the brain is a measure of pleasure, then some schizophrenics ought to be incredibly happy, because their disorder involves the overactivity of dopamine. But that doesn't seem true at all.

Consider the behavior of Russell Weston, the gunman who went on a shooting rampage inside the Capitol building in Washington, D.C., in 1999. Diagnosed with schizophrenia in 1986, according to relatives' descriptions of his behavior in the days prior to his rampage, he showed the classic signs of schizophrenia. He thought, for example, that a gold filling in his tooth was really a radio-tracking device and that his neighbor's TV satellite dish was being used to spy on him. Far from being happy, he was delusional and irritable; in the end, he acted violently because of it. As his parents put it, his mind seemed to "work overtime."

Although Mae West once said that too much of a good thing is wonderful, dopamine is different. There's no doubt it's intimately connected to pleasure,[37] but there's something else going on that is important enough to cause a major shift in the dopamine theory and how we think of ourselves.

During the early 1990s, Wolfram Schultz and his colleagues at the University of Fribourg were trying to understand dopamine's role in movement. Given dopamine's role in novelty seeking, it probably doesn't come as a surprise that it's also involved in movement. You might have seen the movie *Awakenings,* in which Robin Williams portrays Oliver Sacks's struggles in treating patients suffering from Parkinson's disease. Parkinson's involves the destruction of dopamine neurons in a part of the brain called the *substantia nigra pars compacta*. Sacks treated Parkinson's patients with L-dopa, a chemical needed to make dopamine. Patients, like the one played by Robert De Niro in the film, who had been frozen for years suddenly came alive again. They showed new interest in their world and sought its sensations. But the drug soon brought imbalances between dopamine levels and its receptors, triggering manic phases and hallucinations followed by severe depression. This is one of the frustrating parts of developing therapies for brain diseases. The brain's continual shifting to maintain some balance often thwarts a therapeutic drug's effects.

Schultz's team was using very fine electrodes to record signals from single dopamine cells of alert monkeys.[38] Since neurons give off a weak electrical signal when they burst with activity, it is possible to watch the cell's activity on an oscilloscope, like a neural heartbeat. Better yet, you can hear the cell by wiring its electrical signal through an amplifier and a speaker. When the cell fires, the effect is an eerie stilettolike burst of sound. But Schultz and his team were getting frustrated. They had expected to hear the selected cells fire every time the monkey moved, or prepared to move. Yet no matter what they did, they couldn't entice a single burst from these cells. Then one day a researcher casually handed the monkey a piece of apple as a reward during one of the experiments. "The neurons started going crazy. We couldn't believe it," says Schultz.

Schultz and his colleagues performed a series of experiments to rule out that the cell was firing just because the monkey was moving to get the reward. The researchers next considered the possibility that they had found a cell that responds to reward. Had they found a pleasure cell? This is where things started to get intriguing. More experiments revealed that the dopamine cells weren't just responding to reward. Instead, Schultz and his team found that when a light or some other signal reliably came *before* a reward, the neuron would burst with activity. Once this pairing was repeated many times, the neuron no longer fired when the animal received the reward. Instead, it fired with the signal that *predicted* the reward. If you then tricked the cell and didn't give the reward, the cell would dramatically decrease its activity at the expected time of the reward.

Schultz's findings pointed to the possibility that dopamine was involved in making predictions and decision making.[39] If further experiments verify this hypothesis, it would be an amazing demonstration of shared design principles across an enormously diverse range of brains. It would tie together the many other functions of the dopamine system—as both a motivating behavioral facilitation system and the inner fire of positive emotionality and extroversion—with rational thought processes underlying learning and decision making. These faculties would provide you with a stunningly complex internal guidance system underlying central facets of who you are. Indeed, brain science has only begun to scratch the surface of the human internal guidance system. Intertwined with the dopamine system are the serotonin and noradrenaline systems, two other powerful players that have yet to be integrated into an account of the neural basis of our behavior.

At first glance, it might seem naive to claim that a system as ancient as the dopamine one could possibly account for the richness of human decision making. Human decision making is enormously complex. Not only are we capable of pursuing activities that are rewarding in some deep symbolic sense, like seeking mastery in some field, but we are also capable of standing outside of ourselves to monitor how we are doing and can choose actions that do not offer immediate pleasure or reward. Even in the midst of enslavement to drugs, for example, some people can recognize the dangers and check themselves into a rehab center. Upon hearing of an impending increase in the price of cigarettes, many smokers cut back on the number of cigarettes they smoke per day, despite the physical discomfort of their cravings.

Adding the Bells and Whistles

What we've been describing so far might sound more like the behaviorist machinations that Pavlov put his bell-induced drooling dog through than a key to being human. Behaviorism's extreme environmental view ruled American psychology from the early part of the century through the 1950s. It portrayed all animals as basic-vanilla learners, typified by John Watson's challenge that he could take a dozen babies and turn six into lawyers and the others into crooks at will just by altering their environment. The greatest behaviorist icon was B. F. Skinner, famous for his Skinner box and dreams of a behaviorist Utopia that could be made just by skillfully rearranging society's carrots and sticks.

Behaviorism forbade any mention of what might be going on inside someone's head because its leaders like Skinner thought that any talk of beliefs, desires, or feelings was unscientific nonsense that couldn't be measured. Behavior, and its alteration by patterns of reward and punishment, on the other hand, was there for the seeing, and for forty years these adherents to the theory used behavioral yardsticks to measure whatever they could. Thus, the old joke of what two behaviorists do in bed: One turns to the other and says, "I know by your behavior it was good for you, but tell me—was it good for me?"

By the mid-1950s, a new invention appeared that offered an intriguing parallel to what might be going on inside our heads. The invention was the digital computer. Under the impetus of ideas stemming from computers, talk of what was going on inside the head came back into fashion. In particular, the influential linguist Noam Chomsky showed how the computer model of language reshaped thinking about the mind and critiqued Skinner's work on language in a devastating review that is considered a watershed event in theorizing about how the mind works.[40] The mind, according to Chomsky, literally becomes a computer; your thoughts, expressions of your mind's own computer language.

This powerful new way of conceiving the human mind resulted in thirty or so years of highly productive research in cognitive psychology, but it also carried some real costs. We've already seen how applying the computer metaphor results in a seriously distorted view of the mind's development. Whereas behaviorism focused primarily on what was downstairs—what was later found to be the dopamine reward system—cognitive psychology swung the pendulum into the inner world of the mind's computer program. Cognitive psychologists and their spin-off discipline, evolutionary psychology, focus just on the upstairs, the cortex. Disconnecting the brain from the world it inhabits, they believe most everything of interest is built into the brain through evolution, like having dozens of Deep Blues for every specialized ability.

From the perspective of cultural biology, we saw that the pendulum needs to swing back into the middle, building on what is right about both behaviorism and cognitive psychology, and jettisoning the straitjacket ideologies of both camps. Both hold key pieces of the human puzzle. Built-in reward systems engage you with your world and enable you to learn from it, including building symbolic worlds inside your cerebral cortex. In that dual ability is the key to understanding who we are, for it reveals a far more powerful way to make a person than either alternative on its own. The dopamine system reaches up inside the engine of reason, the prefrontal cortex. Reflecting on this, we realized that the real secret to building a human lies in using the ancient predictive systems as an internal guidance system to propel us to engage in our social world, thereby driving the development of complex social cognitive capacities mediated by the prefrontal cortex. Indeed, this powerful interaction operates throughout life, as the dopamine system plays a lifelong role in decision making and working memory. Together, the ancient human core and the evolutionary newcomer, the prefrontal cortex, build the user's guide to life, the self that defines a person.

Building a person this way requires extensive experience in order to tailor the mind's circuits to its world, just as TD-Gammon required experience to build its user's guide to backgammon. This is where the protracted period of human development becomes advantageous.[41] The human capacity to build a user's guide to life, a self that allows us to participate in a complex social world, lies at the heart of human intelligence, affording the mental flexibility that is the hallmark of our species.

As powerful as this system is, we would be returning to outmoded extremes if we suggested that everything that makes you who you are was learned. Bees don't learn to find nectar rewarding. If people had to learn to enjoy sex, you'd have to explain why it would be the most popular course on campus. Some of these preferences are so basic that they've been wired into your brain. Consider Robin Williams's quip that nature gave man the biggest brain and the largest genitals, but only enough blood to use one at a time. He was close. The generator of pelvic thrusts is built right into the spinal cord. Most other built-in preferences and behavioral patterns appear to reside in structures like the hypothalamus,[42] whose output goes to the dopamine system to influence what we find rewarding. As we'll explore in later chapters, these molecules underlie much of our life's texture, perhaps even make love possible. Still, they alone don't dictate who you become. They only provide the fuel that propels the internal guidance system as it interacts with the world to make you who you are.

The end result of this interaction is what we've called a user's guide to life, a repertoire of symbolic selves to help you navigate a complex social world. Your internal guidance system motivates you to seek value all around you, coloring your world with positive rewards and negative punishments. Think about an upcoming event you've been planning for, maybe a vacation, a concert, or even the season opener. For that matter, think of the double chocolate fudge ice cream in your freezer. Take a minute to really think about it. Chances are we can read your mind—you're feeling desire and excitement. Translated into science jargon, the stuff we like, like ice cream, is called incentive stimuli. Seeing it or imagining it propels us into a state of wanting. When you are in this frame of mind, you say you're feeling elated, enthusiastic, peppy, and energetic. These responses reflect the intertwining of emotion and motivation within the human core. Of course, when these responses get overcharged, you can become irritable, nervous, even feel it as butterflies fluttering in your gut. Incentive stimuli have different motivational value. You can measure motivational value roughly by how far you'd walk for something and how far you'd subordinate other goals to get it. For most people, sex and chocolate cake have a lot of motivational value. So it's no surprise that politicians will risk carefully constructed careers for sex, and dieters will throw away a week's worth of calorie counting for the chocolate cake that catches their eye as they pass the bakery counter.

Things ranging from a hug and candy to gambling and sex have motivational value that triggers surges of dopamine in you. Like the monkey whose VTA neurons fire excitedly in anticipation of the slice of apple, your sense of *wanting* something may exceed the pleasure you experience when you get it. Your sensitivity to such experiences depends on biochemical differences between you and others, whether from a genetic or experiential source. Consider a couple we'll call Tim and Debby. On a recent vacation to San Diego, Tim sees the hang gliders soaring above the bluff overlooking the ocean and can't wait to try the sport. Debby's idea of a perfect afternoon, in contrast, is to sit on the beach with the latest bestseller. For dinner, Tim wants to drive an hour up the coast to a restaurant he's heard about. Debby likes the restaurant in their hotel and wouldn't mind going back. Over dinner, Tim talks about how much he's looking forward to trying scuba diving in the underwater caves tomorrow, while Debby's looking forward to finishing her book. Almost everyone knows a couple like this. There's still much debate, but there's good reason to think that the differences between Tim and Debby in sensation seeking are rooted in differences in the dopamine systems in their brains. Tim's genes for dopamine receptors might be slightly different from Debby's.[43] The difference might make Tim's dopamine receptors less efficient at capturing dopamine molecules, requiring greater extremes to have a normal level of dopamine. It also makes Tim seek more novel experiences in search of thrills. Inside your brain, this ancient system likewise underlies your motivation to seek out goals.

We've already induced a feeling of desire in you by asking you to picture the chocolate fudge ice cream in your freezer. But we probably could have gotten you into the same frame of mind just by asking you to think about your ice cream scoop. If you've ever opened your kitchen drawer to put something away and caught a glimpse of your ice cream scoop, chances are it set off a chain reaction in your brain leading to a flash of *Well, just one little bowl wouldn't hurt*. The reason? Whenever you get yourself a bowl of ice cream, you first pull out the scoop. After a while, the ice cream's motivational value rubs off on the ice cream scoop, which acts as a predictive signal for ice cream. The shift of motivational value from primary rewards to any item that is reliably associated with it is a key to brain design because it allows the human brain to discover the predictive structure of the world, as a bee does.

Just as TD-Gammon acquired sophisticated knowledge of backgammon through a back chain of associations, you can over a lifetime acquire extremely complex worlds of value, which underlie everything from why you like the smell of coffee wafting through the house in the morning, to why Michael Jordan's image sells shoes, to why the lingerie business booms. Victoria's real Secret is that the actual style of lingerie doesn't much matter. What does matter is that it's reserved exclusively for intimate encounters, the way your ice cream scoop is reserved for ice cream. If a burlap sack were reserved for intimate encounters, you could elicit the same dopamine effect.[44] The danger is that associating lingerie so strongly with digitally retouched images of Victoria's Secret models might build in the wrong associations, so that eventually real bodies won't trigger as much excitement.

The power of motivational value to rub off on neutral things like ice cream scoops and a few ounces of silk is an indication of the power of this type of learning. The main reason why former drug addicts relapse lies in the nexus of associations surrounding drugs that their brains have built. The mere sight of drug paraphernalia, such as a syringe, or even returning to a site of drug use, can throw a former addict who has been drug-free for years into a cold sweat. Many former smokers get a nearly irresistible urge to light up every time they have a cup of coffee or a meal, events that they've associated with smoking. None of this is lost on advertisers, who spend billions every year trying to build associations in your brain. It's not too surprising that the founder of American behaviorism, John B. Watson, went to work for an advertising firm after he left Johns Hopkins University. As resident psychologist at the J. Walter Thompson advertising agency, Watson devoted his time to promoting Pond's cold cream and Johnson and Johnson's baby powder. His greatest campaign, for Maxwell House, was to make the coffee break an American custom in offices and factories. It makes you wonder whether the factory bell signaling the coffee break wasn't a kind of behaviorist divine inspiration.

Advertising, then, takes advantage of your tireless capacity for on-line learning. This powerful principle of brain design allows you to deal with a world that can't be fully anticipated. Evolution could never have designed you with the engineering principles that went into making Deep Blue because life is far more complex a game than even master-level chess. Instead, nature has designed you to use reward as the basis for continual learning about your world.

Of course, if what you found rewarding was speeding on the freeway or insulting football players in bars, your genes would be at a disadvantage. Rewarding behaviors that enhance fitness tend to stick around, shaping the reward system that gets passed down through the generations. Creatures that found sex and food rewarding tended to be more successful than those who didn't. This is not to say that the brain always has to maximize fitness. It is simply trying to distinguish goals that are rewarding. Some happen to increase your fitness, but some, like drug addiction, don't. The genius of this design is that just by doing what you find rewarding in everyday decisions, you typically wind up in the same neighborhood that you'd be in if you could calculate fitness directly.

Even so, isn't this a sloppy way to go about business? As we've noted, it creates impulses that can get satisfied in ways that have nothing to do with fitness. This is why it is a mistake to try to explain human action solely in terms of maximizing fitness. Drug addiction powerfully illustrates how following rewarding experiences can steer you away from maximizing fitness. There also appears to be a bias in the dopamine system toward favoring novelty, in part because the reward value of a novel situation isn't fully known. It's possible that evolution might have built in a bias to explore these situations to discover whether there's a payoff in pursuing them.

Reward-based strategies place a premium on experience to build a database about the world, the user's guide to life. We were deeply intrigued by how the ancient human core interacts with the evolutionary newcomer, the prefrontal cortex, to build our minds. We now turn to this part of the story.

The Human Story

Antonio Damasio is a neurologist at the University of Iowa and an adjunct professor at the Salk Institute. Nearly twenty years ago, he was asked to examine a patient, whom Damasio refers to with the pseudonym Elliot. In his thirties, Elliot had been a good husband and parent and had a well-paying and respected position with a business firm. But soon headaches made it impossible for him to concentrate. A trip to the doctor revealed a tumor the size of an orange in Elliot's brain. Although the tumor wasn't malignant, its growth was life-threatening. Surgery was performed to remove the tumor and nearby damaged tissue. Later, modern imaging techniques would reveal that part of Elliot's left and much of the right prefrontal cortex were gone, specifically a region known as the orbitofrontal cortex. The frontal regions responsible for movement were intact, as could be seen by Elliot's lack of any motor problems.

Soon after Elliot's operation, it was clear that something very basic inside him had changed. At work, he could no longer finish a task. He would get locked inside a simple decision and spend all day weighing the alternatives—should papers be filed by date, size, or some other criterion? Elliot soon lost his job, and then another. He turned to questionable business deals and got involved with a disreputable figure. Despite the warnings of family and friends, Elliot went ahead and quickly lost everything on a failed deal. Divorce soon followed; then he married a woman his friends disapproved of. Another divorce. By the time Damasio saw him, Elliot was living with a sibling and his medical disability had been cut off.

Damasio recalls that when he first met Elliot he was struck by his charm. He had a good grasp of current affairs and could discuss them with ease. His knowledge of the business world he had once inhabited was strong. He could recall all the events of his life easily. His scores on intelligence tests were either normal or above average. So why had his life fallen apart?

As Damasio spent more time with Elliot, he saw that nothing bothered him. For that matter, nothing seemed to bring him much joy, either. Instead, he recounted the events of his life with a strange detachment. But the strangest change in Elliot and others like him whom Damasio studied was yet to be discovered.

For this part, let's first try an experiment. Conjure up a memory of loss from your childhood. Picture it in your mind for a moment. Now, imagine watching a horrifying image like the sight of the World Trade Center towers falling. Chances are, you not only picture these experiences in your mind but also feel them in your stomach. If we were conducting a real experiment and hooked you up to measure your heart rate and skin conductance with a kind of lie-detector machine, we would see dramatic changes in the movements of the machine's needle. When Damasio and his colleague Daniel Tranel showed emotionally laden images to Elliot, though, the needle didn't move. Elliot no longer had feelings. He could reason about the pictures and describe their content appropriately, but he had no emotional response to them. On other tests, Elliot could reason about moral problems, but he didn't seem able to use that moral insight to guide his own life. As Damasio puts it, although Elliot understood, he couldn't feel.

Damasio had discovered something very important about Elliot and others like him. These patients hadn't lost their social knowledge; rather, they had lost all feeling associated with this knowledge. In particular, they had lost all gut feeling, what Damasio calls "somatic markers," bodily states that mark our experience with strong emotional value.

These cases suggest that without these feelings, we lose our ability to guide our life and cannot choose among future outcomes sensibly. Think about an important decision you face—maybe a move you've been considering, a relationship change, or asking for a raise. As you pictured the scenarios, you probably felt them, too. What you feel depends on your life history and the emotional associations you've built into your life story based on past outcomes. In fact, if we hooked you up to a lie-detector machine and let you go about your daily routine, we'd see many responses that indicate decisions and choices you're not even consciously aware you are making. What makes Elliot's case so tragic is his inability to access the emotional side of his own user's guide to life. Imagine passing through the day or reflecting on your life and not feeling anything. Imagine watching a powerful or romantic movie and having no emotional response. Robbed of feeling, Elliot can't use his knowledge to predict what would happen if he took certain actions. That is the essence of making decisions that affect our life: weighing different options, imagining different worlds, and choosing action. As Elliot himself recognizes, he cannot even learn from his mistakes.

So severe is Elliot's loss that Damasio calls his condition "acquired sociopathy." As we'll explore in chapter 8, an even more terrifying possibility lurks inside the mind of a developmental sociopath, particularly the likes of a John Gacy, Ted Bundy, or Jeffrey Dahmer, one whose early life experiences and predispositions combine to stunt the capacity to feel and care about another person.

Elliot's damage was to his prefrontal cortex. We've mentioned that the dopamine system sends many of its fibers to the prefrontal cortex, hooking an ancient structure to an evolutionary newcomer closely associated with human qualities. This linkage is remarkable. Researchers aren't entirely clear on what the conversation between the two sounds like, but the links between decision making and dopamine suggest that it may play a crucial role in our highest forms of thought. We've noted that the bizarre and confused thinking that is the hallmark of schizophrenia, a disorder involving both the prefrontal cortex and dopamine, gives further credence to this possibility. All this suggests that the cardinal distinction between thinking and feeling has been misleading, and that feeling may be more important in our everyday lives than we imagine.

Of Human Goals and a Value-Added World

One of the century's most influential psychologists, Abraham Maslow, suggested that there exists in us a hierarchy of needs. At the bottom are needs like sustenance, warmth, and shelter. At the top are needs such as that for self-actualization. Whether or not you agree with the details of Maslow's scheme, the idea that humans have different needs, from basic ones to higher ones, is compelling and may explain the existence of religions in every culture, as religions address human needs at every level.[45] What you find rewarding depends heavily on the worlds of meaning you and others construct collectively in a society. You do some things on the basis of what you believe, beliefs that hold deep value for you. This is the definition of an ideology: a belief system that you defend against competitive beliefs because it holds such deep value.

On reflection, there is something bizarre about how much human history is a clash over conflicting beliefs. There has been almost no progress in resolving some of our most intractable conflicts today, such as those in Northern Ireland and the Middle East. Perhaps we have thought about these clashes of ideology in the wrong way. Can beliefs be addictive? Just as drugs can be rewarding in ways that have nothing to do with fitness, so too belief systems can be addicting, both in ways that enhance fitness and in ways that make you go astray. It is no accident that repressive regimes throughout history have most feared the power of ideas.

Perhaps a drug model of beliefs would be more revealing than a rational account. Consider the tempest that a belief can spark inside the brain. Unlike the cold-bloodedness that makes a killer out of a sociopath, it is obsessive ideation that can produce misfits like Timothy McVeigh. When McVeigh was arrested he was wearing a T-shirt depicting a tree with drops of blood for leaves. Under it was Jefferson's remark, "The tree of liberty must be refreshed from time to time with the blood of patriots and tyrants." On the back was an image of Lincoln in a wanted poster. After he assassinated Lincoln, Booth yelled, "Thus always to tyrants." What sort of world did McVeigh and his fellow self-styled patriots inhabit? What kind of values did it instill in them that led them to justify their actions as those of heroes, a belief that McVeigh held until his death?

The emerging view of who we are is providing glimpses of a human core that is far different from the model of a brain rigidly wired by evolution. Instead, experience, thought, and feeling intertwine as we experience the world and build our sense of self and our user's guide to life. This dynamic interplay fosters the unparalleled flexibility that marks each of us as unique. What is most surprising is that the making of a human rests on systems that are at least five hundred million years old, intricately tangled inside an expanded brain. It is this tangled web that helps drive the construction of your self. But how does it come into being?

Becoming You 6
You Genes,
Parenting, and
Personality

Although the temperature on the tarmac is over one hundred degrees Fahrenheit, it hasn't deterred more than a hundred thousand people from showing up. Suddenly, the silence is broken by two F-18s screaming in from opposite directions a hundred feet above the ground. In a game of chicken played just below the sound barrier, they converge on each other at a closing speed of over a thousand miles an hour. Just as they are going to collide, a last-second snap on the sticks rolls each plane ninety degrees, narrowly averting disaster. A second after they pass one another, both planes enter a maximum turn, pushing their pilots into their seats as though they weighed a thousand pounds. Then both pilots hit their afterburners, spilling raw fuel into their jet's exhaust port. The fuel's explosion turns their engines into rockets, and both planes ride its crackling power straight up to disappear from sight. A moment later they reappear, converging on one another again. Except this time they are flying upside down.

The site is the annual air show at air station Miramar in San Diego. The enormous sign over a hangar reveals its unofficial name, Fightertown USA. It is famous as the place where Tom Cruise came to scratch his itch for speed at its Top Gun flight school.[1] The pilots wowing the crowd today belong to the Blue Angels, the U.S. Navy's high-performance team. For them it's just another day at the office, pushing the flight profiles of their planes to the utmost while less than a yard from each other. Fighter pilots have the right stuff, a fearlessness and boldness that make them bored unless they are pushing the envelope. And among fighter pilots, there's no cooler customer than John Glenn. The veteran of 149 combat missions in World War II and Korea, Glenn's idea of fun was sneaking up on other pilots flying in his formation, including baseball great Ted Williams, and nudging their wingtips with his, all while cruising at six hundred miles an hour. Sitting atop the Saturn rocket that lifted him into space as the first American, Glenn's pulse never went above an icy 110 beats a minute, not even jogging pace. At the age of seventy-seven, when many other people's idea of a thrill evokes images of golf carts and an early tee time, Glenn took the second trip into space he'd been trying to convince NASA to give him for some thirty years.

Are fighter pilots like John Glenn made or born? With some coaxing, would it be possible to turn a Woody Allen into the next wingman for the Blue Angels? What makes us individuals? Are we born a certain way, not to veer from what is written in our genes except by the most extreme environmental forces? Or does our early experience indelibly shape who we will become? If so, what are the crucial experiences and forces that shape us? Did Glenn's parents toss him into the air as a child and expose him to thrills? Did Allen's parents fret over him and overprotect him? While Glenn was out racing cars, was Allen forced inside to practice the clarinet?

These are among the questions we probe in this chapter. They will take us into the heart of the human mystery and the first suggestive hints that cultural biology holds for the mystery of human individuality.

A Starting Point: Genes and the Baby

It's becoming a cliché to say that parents are environmentalists until they have their second child. One newborn may spend his awake time quietly inspecting his surroundings with the calm of a Buddhist monk, while another may spend it screaming inconsolably. Despite what parents of placid children may secretly believe, it is well established that we are born with these basic dispositions. Some of the best evidence for basic dispositions comes from Harvard child psychologist Jerome Kagan. Kagan, who describes himself as a politically liberal social scientist trained to believe in the environment's extraordinary power, reluctantly came around to the notion of basic dispositions after fifteen years of research. Kagan studies temperament in newborns and small children; that is, their characteristic mood, activity level, and style of responding to the world.[2] According to Kagan, by four months of age, about 20 percent of infants have the behavioral and physiological signs of an inhibited temperament. In contrast, about 40 percent of infants have signs of a bold or fearless nature, with the rest somewhere in between. The differences can be striking. For example, a loud bang might spark the shy infant into a bout of tears, whereas the bold infant might interpret the same bang as an invitation to explore something interesting. This is the essence of temperamental differences: Individuals assign emotions to experiences very differently, and so interpret the world in their own characteristic way.

Since babies haven't had much experience with the world,[3] their styles of temperament likely reflect some basic differences in neural and biochemical machinery. There is a deep connection between your internal guidance system and your temperament: Since the internal guidance system shapes infant behavior, temperament most likely reflects the configuration of this early behavioral system. This connec-

tion is supported by the fact that this system helps color adult personality, from extroversion to positive moods.

Although the neural underpinnings of differences in temperament
are still far from being fully understood, it's clear that noradrenaline,
serotonin, dopamine, acetylcholine, and their interactions are all involved. As we noted in chapter 5, these internal guidance systems stem
from small clusters of cells deep inside the brain that send their axons
outward, spreading across most of the brain, modulating the responses
of other cells.[4] Relatively small changes in these systems' chemistry reverberate throughout most of the brain. Tweaking a few key genes in
these systems would thus be a fairly easy way to make appreciable
changes. For example, children with high levels of noradrenaline react
strongly to low levels of stimulation, and children with low levels of
noradrenaline have low levels of arousal and so may require more
stimulation. It is likely that differences in temperament are the result
of long-term alterations in levels of noradrenaline, serotonin, and
dopamine, as well as in their complex interactions.

The set points for your neuromodulatory systems also depend on
experience. Indeed, the growing evidence that prenatal experience
plays a role in shaping the internal guidance system's neurochemistry
is fascinating. In particular, baby monkeys whose mothers are exposed
to a major stress while pregnant show the signs of that stress in their
temperament.[5] These babies spent more time clinging to their caregivers and less time playing with cage mates, displaying an inhibited
temperament reflected in their altered levels of noradrenaline and
dopamine. These infants also had difficulty forming attachments to
their mothers, which in turn resulted in more adverse experiences,
causing behavior to go from bad to worse.

Despite the role of early experience, genes set the stage in shaping
temperament. If changes in one or a few genes could have potentially
large effects on temperament,[6] we thought, how much does genetic
variation underlie differences in temperament? Does anything stop
variations, known as polymorphisms,[7] in the genes involved in these
systems?

We've noted that much of your genetic endowment is highly conserved, meaning that you share it not only with fellow humans but also with many other species. In fact, you share 99.9 percent of your DNA with everyone else on the planet—we are all closely related. All human diversity flows from the remaining 0.1 percent. There's a good reason for this conservation: In most cases, tinkering is deadly. In humans, genetic mutations leading to deformed heart valves, for example, can cause such a drop in fitness for their owner that they aren't likely to get passed on. Most polymorphisms occur in so-called junk DNA, which doesn't contain recipes for making the proteins we need for life. Mutations in junk DNA probably have less bearing on the immediate fitness of an individual than mutations in coding regions.[8]

Considering the brain's remarkable complexity, it's amazing that congenital brain defects comprise only about 3 percent of births. One reason is because the chances of carrying a fetus with major brain defects to term is similarly low. Fully 75 percent of miscarriages are due to brain malformations. For those that do survive to term, 40 percent of infant mortalities in the first year are due to brain defects. On the other hand, if a polymorphism in a gene produced a large fitness increase, then, like a good idea, it would tend to spread quickly through a population. Maybe differences in temperament don't confer any large advantages. Analogously, that's why the world is filled with tall and short people, since neither bestows substantially more fitness.

In that context, let's return to our comparison of Woody Allen with John Glenn, paradigms of the inhibited and fearless poles. In evolutionary terms, who is more fit? Despite their enormous differences, they both managed to find a place in the world. And although Glenn's macho fighter pilot image might have brought the women flocking, racing cars as a kid and flying bombing runs as an adult isn't the best prescription for living a long life. In contrast, while Allen's nervous and timid exterior might not make for the most effective personal ad (waifish neurotic man seeks . . .), it hasn't been an insurmountable obstacle for reproduction opportunities. This suggests that the different genes underlying temperament have reached an evolutionary stalemate.

One might even be tempted to argue that society needs diverse personalities in order to function, but this line of reasoning is questionable because it assumes that natural selection works for the good of the species. Instead, we think that the diversity of temperaments in individuals reflects the variety of different niches in society for which those temperaments provide survival advantage.

Does Being Born *Some* Way
Equal Being Born *That* Way?

The existence of basic temperament styles dispels the notion that newborns are empty, passive vessels into which the world is poured. Although you were born helpless, your head flopping under its own weight, you were born with your internal guidance system activated: You were ready to explore your world. You recognized the sound of your mother's voice, which you had listened to even in the womb. Faces, which babies prefer to other objects, transfixed you. Your early temperament also helps your parents to want to care for you, to form the bonds of attachment that your brain is primed to thrive on. Thus, babies are born knowing how to focus on the parts of the world that are essential for learning about how to navigate in it.[9]

But doesn't the fact that we are all born some particular way sound depressingly deterministic? No, because being born *some* way doesn't amount to being forever destined to remain *that* way. Genes help determine your height at birth, but your environment—for example, what you eat as you mature—also has an important influence. The average European today is eight inches taller than the typical European 150 years ago, a change that is attributed to environmental factors.[10] So it is with temperament.

By four years of age, the results of a gene-environment interplay are evident. We mentioned that about 20 percent of four-month-olds have an inhibited temperament, while about 40 percent have a bold or fearless nature. In contrast, only about 10 percent of four-year-olds belong to these extremes. Something has intervened to temper temperament.

A clue to what that something might be can be found in the cages of the National Institutes of Health Animal Center in Poolesville, Maryland, where Stephen Suomi's fifty-plus young rhesus macaques are providing intriguing clues into the modifiability of early temperament. Suomi, a primatologist at the National Institute of Child Health and Human Development, selectively bred monkeys for both inhibited and bold temperaments. Then, he rearranged their rearing environments to explore how environmental factors and temperament interact, all while tracking their neurotransmitter levels.[11] In Suomi's experiments, a fearful infant monkey was put in the care of an uninhibited, nurturing foster mother. Suomi found that over time the infant monkey's behavior and biochemistry came to reflect those of the foster mother: The young monkey became less fearful, and its levels of noradrenaline dropped. Regarding temperament, Suomi notes, "Our work shows that you can modify these tendencies quite dramatically with certain types of early experiences."[12] Like our research that demonstrates how experience can help build the circuits of thought, this work shows that it also helps build the circuits of emotion and personality. As yet, however, a systematic understanding of how to tailor individual environments to help each human flourish remains largely unexplored. Such research remains a valuable, if elusive, goal whose implications we will revisit in later chapters.

From Temperament to Personality

Temperament colors you in broad brush strokes. Mediated by your internal guidance system, temperament shapes how you respond to the world and the people in it. Those responses in turn create new experiences, setting in motion a cascade of formative experiences. This isn't the whole story, though. We haven't yet touched on your user's guide to life, which is mediated by prefrontal brain structures. As you develop, behavioral control shifts from your internal guidance system to include your user's guide to life, a shift sometimes called the frontalization of behavior. The same shift occurs with temperament: Whereas temperament is the main component of your internal guidance system during infancy, your personality unfolds slowly over time as your user's guide to life comes online. As we will discover, it takes a long time to build your personality and the flexible behavior it allows.

Is personality just a more complex version of temperament? Thomas Bouchard and other behavior geneticists think so. They describe personality as a largely unmodifiable behavioral system, citing research on twins separated at birth, who seem to have strikingly similar personalities.[13] This is also the view of sociobiologists, who see personality as an innate behavioral system with traits that have been selected for their fitness. The views of many evolutionary psychologists are a variation on this theme, with personality defined as a set of instincts for solving social problems.

The alternative to preprogrammed instincts is what evolutionary biologist Ernst Mayr calls an "open program." As we saw in the last chapter, Deep Blue and TD-Gammon are, respectively, examples of an instincts program and an open program. Deep Blue comes to the table with its chess knowledge built in. TD-Gammon, on the other hand, utilizes an open program to develop game knowledge as it plays. To do this, TD-Gammon needs a lot of practice, relying on flexibility, an important component of intelligence, to discover solutions to unanticipated problems as they arise.

Is human personality an open program? We think so, and for good reasons. Open programs have the advantage of shaping personality for changing and unanticipated roles, or niches. They can even help create new niches. This powerful feature is the hallmark of the human mental flexibility at the core of cultural biology. Slowly, as you experience your world over many years, out of the partnership between your anchoring internal guidance system and your user's guide to life emerges an intellectual coup: the impressive flexibility that allows you both to reshape your world and to respond to its personality-shaping contexts. Indeed, although children essentially master cognitive skills such as language by age four or five, human development continues through adolescence because it is an enormous task to build a user's guide to navigate a complex world. The process of building this guide is called socialization, and it is a long road that you must travel. Much of what is called personality is formed along that road.

On the long road to personality, what brain region might be at the wheel, guiding your interaction with complex environments? A leading candidate is an intriguing brain region called the anterior cingulate cortex, which we discussed in chapter 2. The anterior cingulate is part of a group of brain areas termed the limbic system, which uses your previous experience to organize appropriate cognitive and emotional responses in new situations. The anterior cingulate is often active in functional imaging studies when a person is trying to solve a highly demanding problem.[14] Recall that damage to the anterior part of the cingulate cortex leads to a condition known as akinetic mutism, immobility and impaired interaction with the world.

At Caltech, John Allman has discovered that a type of neuron in the anterior cingulate of great apes, called a spindle cell, is numerous in humans, less common in chimpanzees, rare in other great apes, and absent in all other mammalian species.[15] This cell type is not present in the human cortex at birth, but only appears at four months.[16] Spindle cells, which project to many parts of your brain, could carry information to help monitor your online performance when you solve difficult problems. There is mounting evidence from a wide range of studies that your anterior cingulate may be especially important for your ability to concentrate and focus your behavior, especially in social settings, so the appearance of spindle cells at four months suggests that there is a direct link between widening social competence and an emerging user's guide to life.

Enter Freud

The long postnatal development of human spindle cells illustrates the collaborative project between experiences and the world that builds personality. The classic account of this collaborative project is Freud's, but he preferred to focus on the dysfunctional possibilities lurking in open programs. Early in his career, listening to his patients recount the sexual abuse they experienced as children, Freud theorized that the process of personality building could be derailed by infantile seduction. Later, however, he decided that these patients weren't recounting actual events of sexual abuse at all, but rather memories of wishes and longings. Freud placed these sexual wishes in a sexual instinct: the libidinous energy of the id. His theory culminated in his formulation of the Oedipus complex, which energized the child.[17] With the child's wants created by this drive, Freud believed that the parent-child relationship was one of conflict, whose early battles shape personality.

Freud's proposal is still deeply ingrained in us. For example, when asked about her husband's errant ways, Hillary Clinton suggested that they had their roots in his tumultuous relationship with his mother and grandmother as a four-year-old, an answer that stems from the quintessentially modern idea that your early childhood relationships with your parents make you who you are. Indeed, from the daily fare of talk shows to shelves of self-help books, from pundits analyzing the murderous impulses behind tragedies like the shootings at Columbine High School to Dr. Drew's analyses of sexual dysfunction on MTV's *Loveline,* the idea that parental relations decidedly shape your personality is a cornerstone of the modern identity.

Over the past few years, brain science has been recruited to support this essentially Freudian notion. In special issues of *Time* and *Newsweek,* in the 1997 White House Conference on Early Childhood Development and Learning, and in countless other forums, the argument has been made that the first few years of your life offer a narrow window of opportunity during which the brain rapidly forms nerve connections, laying down the circuits for everything from acquiring a second language to musical ability to a healthy personality. According to this view, after your first few plentiful years of cerebral life, your brain begins a long process of paring down its exuberant connections, just as a sculptor works a piece of marble. Traumatic events during early childhood could disrupt a critical period of growth, making the adult brain, in effect, a flawed, unworkable piece of stone.[18]

Although this view remains immensely popular, and the toxic effects of early abuse and stress are central to our exploration of violence in chapter 8, the new themes of brain development we explored in chapter 3 suggest that the formation of personality needs to be rethought, with far-reaching implications for understanding how you become who you are. Since the circuits of personality take two decades to be built, far beyond your "formative" years, the driving force of personality development must be more than just your early interactions with your parents. What are the forces that shape you?

Judith Rich Harris, a psychology textbook writer and self-styled maverick disguised as a mild-mannered grandmother, suggested an answer in her much-discussed, widely vilified book *The Nurture Assumption,* in which she dared to argue that parents have no long-term effects on their children's personality, intelligence, or mental health. Instead, she believes a different personal history matters: that of peer groups. To her, we are shaped by our efforts to find our place in the larger contexts of our social world, outside our family's influence. In fact, Harris believes that given the same community children would turn out just the same even if they were all swapped at birth.

Harris ultimately sides with those who view personality development as an open program. That is, personality isn't inborn; it's a survival strategy that gradually unfolds to navigate the particular slings and arrows of your situation. This much Harris shares with Freud. Yet Harris's idea veers from a Freudian perspective, for she claims that personality unfolds gradually from infancy to adulthood—a claim that resonates deeply with recent findings in brain development. We, too, were drawn to the idea that personality unfolds gradually over time, but we questioned whether it meant we also had to abandon the notion that your early experiences play a formative role in personality development.

MIT's Frank Sulloway would argue that Harris has literally thrown the baby out with the bathwater. Sulloway views the household as a kind of Darwinian theater with the parents as directors and children auditioning for their roles. To see what these dynamics might look like, we return to Darwin's finches of the Galápagos. Although the thirteen finch species bore a striking resemblance to one another and to those on the mainland, there were some major differences, particularly a remarkable diversity of bill shape and size. Darwin speculated that these thirteen species likely started as one species, perhaps as a few finches blown to the islands from the mainland. Over the generations, they changed into the thirteen distinct species. The question was, why? In a process called adaptive radiation, new finch species with their own distinct bill shape and size would be capable of foraging on foodstuffs that other finch species couldn't utilize, allowing a new finch species to survive in an unoccupied ecological niche.

Sulloway suggests that an analogous process is at work inside the Darwinian theater of the family, but the real drama isn't a Freudian parent-offspring conflict. It is the competition among siblings for a share of a limited resource: parental investment. The firstborn has a natural advantage, arriving early for a private audition. For that reason, firstborns tend to be a bit like stage managers, wielding their "in" with the directors to whatever advantage they can muster. Scouring historical records and analyzing birth-order studies, Sulloway concludes that firstborns tend to be more concerned with achievement, more ambitious, angry, antagonistic, anxious, assertive, conventional, deferential to authority, dominating, identified with parents, jealous, and self-confident. They also tend to be less empathetic, less willing to identify with the underdog, less innovative and open to learning by experience, less rebellious, and less willing to take risks. As a result, firstborns are overrepresented among American presidents, British prime ministers, and participants in less risky sports such as swimming and golf.[19]

According to Sulloway, in this family scenario, a laterborn faced with a bossy older sibling has a couple of alternative strategies. One is to go head-to-head with that sibling, essentially competing for the same role in the play. Most siblings, however, are more like Darwin's finches: They seek out their own niche within the family ecology. They try to differentiate themselves, and in the process make themselves valuable within the family. The difference is, in families it's not beaks but personalities that diverge.

This is a psychodynamic theory, like Freud's, meaning that personality isn't just some canned set of instincts, but instead a flexible and adaptive set of strategies for fitting unanticipated niches. Your DNA can't anticipate what order of birth it will find itself in when you are flung into the world, so it must create open programs. Indeed, to Sulloway it is primarily the laterborn's openness to experience, the capacity to seek out and incorporate new experiences into behavioral systems, that drives this Darwinian-like divergence. In Sulloway's words, "Personality is the repertoire of strategies that each individual develops in an effort to survive childhood."[20]

Sulloway acknowledges that the Darwinian theater has more structure than just birth order. Additional influences on the finches' adaptive radiation included the productivity of the islands, the habitat diversity, and the population density of the plants and animals involved. So, too, birth order would have to be understood alongside other variables, such as available parental resources, family size, gender, temperament, class, and so on. But the crux of Sulloway's argument is that among these variables, birth order has the largest effect.

So we have uncovered two opposing theories of personality. One points to the peer group; the other points to a Darwinian drama within the family. When confronted with this polarized, "either/or" debate, it's always useful to ask if the debate itself rests on false dichotomies. Perhaps there is something fundamentally askew in trying to isolate a single context as *the* driving force of human personality.

The Trouble with Traits

We realized that Harris's and Sulloway's basic mistake was to view human personality formation as driven by the search for a single, unoccupied niche within either the family or the peer group. Your personality—your user's guide to life—makes you supremely flexible, able to navigate a variety of niches—from the home to work and social events—that you encounter throughout the day. In more than a metaphorical sense, you aren't the same person at home as you are in the office. This inconsistency isn't a defect. Your capacity to gauge context and adjust your personality accordingly is a remarkable achievement that is at the core of cultural biology. It is a key to human social

life. It underlies our ability to inhabit many social worlds, which sometimes takes ominous form (see Philip Zimbardo's prison experiments in
chapter 1).

The idea that personality is a flexible behavioral repertoire doesn't
fit with dominant theories of personality today, so-called trait theories,
which both Harris and Sulloway endorse.[21] According to trait theory,
your personality remains stable across many different types of situations and predisposes you to certain behaviors. The most popular trait
models of personality are "big-five models." Both Sulloway and Harris
view personality as a combination of five essential traits: perhaps a
drop of extroversion, a dash of agreeableness, a dollop of conscientiousness, and a light dusting of neuroticism, topped with a sprinkle of
openness to experience. Putting them all together, you get a characteristic style of responding to different situations. According to this
model, Sulloway suggests that you dialed in different levels of these
five traits that together adapt you to a niche within your family. For example, as a typical firstborn you may be extroverted, antagonistic, conscientious, jealous, and conservative. Harris, on the other hand, argues
that your traits were selected by your efforts to find a niche within your
peer group.

Trait-model theorists have long tried to confirm their view by exploring how much "cross-situational consistency" there is in human
behavior. Their research, however, has had disappointing returns.
Human behavior shows only very modest consistency across situations. It turns out that you adjust your behavior to fit specific contexts,
making you more context-sensitive than consistent. Some trait theorists have tried to patch up the flaws in trait theory by citing the law of
averages. Perhaps you're not very conscientious on a Friday night,
when you work the bar scene, have one too many, and decide to drive
yourself home. At the office the following Monday, though, you're exceedingly conscientious as you work on a report. If we average your behavior, you end up scoring pretty well for conscientiousness. The
trouble is, you aren't an average. What we'd like to know, and maybe
your friends would, too, is why you acted like a jerk on Friday night.
The averaging approach to personalities tosses out what is most dis-

tinctive about you and omits the most important factor: the context sensitivity of your behavior.[22]

This personality paradox has split the field. Some psychologists have attempted to replace the trait model with "process" models of personality that acknowledge the shaping power of context. Among the most influential is that presented by Walter Mischel and Yuichi Shoda, who assert that our personality must be thought of as a context-sensitive behavioral system.[23] In their words, this capacity is a "basic aspect of social competence, not a reflection of inconsistency."[24]

This is consonant with the key claim of cultural biology that mental flexibility is the hallmark of intelligence. Our greatest evolutionary achievement meant escaping the niche limitations inherent in acting the same in different situations. In fact, there are few social embarrassments more painful than watching a friend who can't read social cues and acts inappropriately. Imagine if you acted in front of your boss the way you do while relaxing at home or on the weekend with friends. The fact that you can behave appropriately in such strikingly different contexts reveals just how limber your brain is. That's why human personality isn't like Darwin's finches. We need to widen our view of human behavior, embracing our context sensitivity as a hallmark of who we are.

That's the upside to human mental flexibility. As we will explore in later chapters, this flexibility also comes at a cost. As much as we would like to believe that life's decisions are a matter of personal character, the power of context is often overpowering. If you think that there is an inviolable core within you that dictates how you behave despite any context—something called character—a growing mound of psychological and historical evidence suggests that you may simply have never been exposed to an extreme context in which to test yourself. Character may be an essential—but largely baseless—story we tell ourselves.

From Personality to the Neural Self

Returning to the question of what drives personality formation, since the essence of socialization is learning how to live among a multitude of contexts, there is no reason to choose between the family environment and the peer group. Personality formation is subtler than that. The fact that children don't behave at home the way they do with their peers doesn't mean that parents have no influence over their children. It only means that children are learning how to live in more than one context. After all, why should children behave with their peers the way they do at home? The bossy firstborn isn't going to get too far with that strategy if her peer group contains a more dominating figure.

As children mature, their worlds broaden, deepen, and differentiate. Their experience begins with the family context, then expands to the world of child care and school, peer groups, neighborhoods, and eventually to the workplace and other extracurricular settings.[25] Children first engage in these contexts to satisfy basic needs, but over time they create more intricate relationships with others to satisfy more complex psychological needs, such as those exemplified in the tangled webs of adolescent friendship.

The Sense of Being You

As children enter increasingly complex relationships, they are not only developing personality but also constructing their identity. As William James, the great pioneer of American psychology at Harvard and brother of novelist Henry James, noted over a century ago, thoughts don't just float through your mind, they belong to you. James called this experience of owning your thoughts the personal self, the *I* that anchors your thoughts to you. This capacity to identify your experiences as your own allows you to weave a story from personal memories, with you as the leading character, extending both back and forward in time.

Your capacity to organize your experiences around an enduring subject separates you from even your closest genetic cousins. This is the essential cognitive capacity that underlies human social life. It allows you to time travel mentally, to recall past scenarios and construct future ones, thus allowing you to form long-term plans. Perhaps most important, just as you can understand yourself as a person with goals, hopes, desires, and beliefs, so too can you see others as persons of like mind. Just as theories from physics allow us to navigate the solar system, theories about minds allow us to navigate the complex social worlds we inhabit.

This crucial piece of our user's guide to life is frequently called a folk psychology, since sometimes we learn it on Grandma's knee as she tells us about the ways of the world. Only humans interact with each other by using a folk psychology. Chimpanzees, in contrast, appear to be mind-blind, unable to conceive of their cohorts as having minds. With these capacities to see yourself and others and to plan a future based on current and past contexts, you can construct symbolic worlds of culture that contain a moral order, rules of conduct, beliefs about history, and visions of a future.

The story of the Sebei girl in chapter 4 illustrated the profound implications of this social capacity. The world of self and other adds a moral order to group life. This symbolic world of roles, obligations, and expectations is a deeply complex game that you only began to master in adolescence, self-regulating your behavior in accord with the prescriptions of society and with the internal models of self that you have constructed. Key to regulating your behavior is your capacity to see into the shadow of the future, envisioning it and acting in a way to bring it about.

How a three-pound system of cells somehow gives rise to self-reflection and the sense of an enduring personal identity is one of the greatest scientific mysteries left to be solved. We certainly don't want to give the impression that we are going to explain it here, but brain science has now reached the point where we can begin to frame the questions in ways that can be answered, a remarkable achievement in its own right. In the light of recent advances in brain science, we can explore where our idea of a self came from and where inside the brain it may reside.

Consider the case of an amnesic patient known as M. L.[26] In June 1993, M. L. was an active man, a successful sales representative with a two-year-old daughter and a son on the way. M. L.'s life was shattered while out on a bike ride. Struck by a fast-moving car, M. L. hit the ground hard, sending him into a coma. Brain scans revealed a subdural hematoma, a mass of blood in the space between the brain and skull. When M. L. regained consciousness six days later in the hospital, he no longer recognized his family. Today, he still remembers only a few disjointed fragments of his past.

What is particularly strange about M. L.'s case is that he can learn new facts and tests normal on most recall and recognition tasks, but he can't recall events from his past, even those that occurred after his accident. In other words, he can learn and recall facts, but he can't integrate them within the fabric of a personal self. M. L.'s experiences have become unhinged from their owner. When M. L. was asked to recall events from his personal past, images of his brain revealed that his right prefrontal cortex, an area believed to be involved in recalling personal memories, barely activated at all. It is intriguing to note that this diminished activity occurs in area 10, a region of the prefrontal cortex that is substantially larger in humans than in chimpanzees. This convergence of evidence suggests that this area might underlie our capacity for self-awareness.

M. L. shows us why having a sense of who you are is so crucial: Normal, everyday situations are inherently ambiguous in that the appropriate path one should take cannot be determined solely from the environment. Having a sense of self eliminates this ambiguity by allowing you to call on your personal history and long-term goals to formulate an appropriate strategy. Since M. L. can't form goals based on his own identity, he is forced to regulate his behavior by relying on generic information about how one should behave; it is like he is perpetually on stage without a script. As a result, M. L.'s life is marked by a self-regulatory disorder, from an inability to regulate his own behavior to an inability to safely supervise his children.

Inventing Yourself

In chapter 4, we noted that chimpanzees appear to have one critical component of culture, namely, the ability to transmit behaviors socially. Yet no matter how long you watch a group of chimpanzees in the wild, you never see one hold up an object to show to others or intentionally teach another a new behavior.[27] The same is true of human infants until around nine months of age, when something revolutionary happens. They begin to draw adults into shared interactions with other objects. Steve recalls vividly when his daughter, Evelyn, then around ten months of age, first gestured for a stuffed bear that was out of her reach. For the first time in her life, she somehow understood that she could direct his attention to the object she wanted and thought of him as an agent that could help fulfill her goals—a lesson that she and her siblings continue to make good use of to this day.

As Evelyn accumulated more and more experiences of herself as an entity that could act on the world, she developed a budding sense of self that served as the foundation for seeing others as like herself. In other words, her world began to include objects with very special properties, namely, people. Almost all parents are relieved when their children reach this stage, when they become little people and take delight in engaging with others this way. It allows a new form of cultural learning based on the ability to think about self and others as having goals.

Nonetheless, very young children are extremely limited in the kind of social exchanges they can engage in. Their rudimentary sense of self may be linked to something else they lack: language. Indeed, it is intriguing to note that children under the age of two are like M. L. in that they have no autobiographical memories. Typically, children are around three and a half years old when they acquire autobiographical memories. Infantile amnesia is overcome when children learn from their parents how to form their memories into narratives, with language acquisition a crucial part of the story. In particular, learning how to use personal pronouns provides children with a way to think and talk about their personal self. As children interact with others linguistically, they elaborate on their personal self in a process that continues through adolescence and indeed across their life span. By the time they are about ten years of age, children are comparing themselves with others for self-evaluation and are internalizing others' opinions and standards, a critical milestone on the road to the mature person.

As children's sense of self deepens, so does their sense of others. A critical event occurs around four years of age, as was recently seen by Steve. At the time, his sons were part of a play group that included an autistic boy who lived down the street. Elliot, then a few months shy of four, and Alden, six, were curious about what autism was. Steve demonstrated by performing an experiment that has been used extensively with autistic children. Sitting around the kitchen table with Steve's wife, Karen, Steve took out a candy bar and, in front of the boys, hid it inside a drawer. Steve then asked Karen to wait in the other room and come back in a few minutes to retrieve the candy bar. When she had left the room, Steve moved the candy bar to another drawer. He then asked Elliot where he thought Karen would look for the bar when she came back in. Elliot said she'd look for it in the second drawer. Immediately, Alden spoke up and said that was wrong. Karen would look for it in the first drawer.

Elliot's answer revealed that he could not yet attribute a false belief to another person. He knew the candy bar was in the second drawer because he saw it moved there, but he couldn't distinguish between his own beliefs and his mother's false belief about the candy bar (namely that it would be in the first drawer). As Steve explained to them, Elliot wasn't able yet to take the perspective of another person, which is a critical element of theory of mind. Alden, in contrast, could reason about what his mother would do even when it meant attributing beliefs to her that he knew to be false. Their autistic friend might never be able to do this, or he would have other difficulties in reasoning about the minds of others.

Your ability to think about yourself and others is mediated by the prefrontal cortex,[28] whose long development partly reflects the many years it takes to construct sufficiently complex theories of self and others to navigate a complex social world. What is really intriguing is that the construction of the self depends critically on social interaction, in-

cluding linguistic exchanges. We can see this dependence in deaf children, who are typically born to hearing parents who have no experience
in sign language. Because the parents and siblings of deaf children
must learn how to communicate with one another through sign language, linguistic exchanges are typically simpler than those in a hearing family. For this reason, deaf children are often delayed in their
mind-reading skills and in the development of their self-concept.[29]

If your self is socially constructed, and if social competence demands
that you construct many selves for differing contexts, what implications
does this have for our cherished notions that self-improvement entails a
search for an authentic, inner self and that our character acts as an internal moral compass?

Many of these beliefs are the result of such entrenched historical
traditions that we accept them without much reflection. Many in the
West, for example, live in a deeply individualistic tradition of the self.
This is probably one reason why Judith Rich Harris's emphasis on the
group received such disapproval. Harris's 1995 *Psychological Review*
article ended with the African saying "It takes a village to raise a child,"
which Hillary Clinton used for her 1996 book on child development.
Only against a long tradition of individualism could that become a political cry to arms.[30] But this individualistic conception of the self is a
historical invention.[31]

Renaissance humanism put the "sovereign individual" at the center
of the universe; later, the Reformation liberated the individual conscience. This new conception of the individual was the cornerstone of
liberal democracy, which stemmed from Thomas Hobbes's and John
Locke's treatment of the individual as the fundamental unit of society,
an idea we'll explore in more detail in the next chapter. In their view,
the individual no longer lived for the good of society and did not require society for a sense of personal identity. Whereas before the individual had existed to promote the good of society, now society existed
for the good of the individual, to enhance personal liberty in what is
known as the priority of the right over the good.

The sovereign individual was the engine of modernity, the rational
inquirer in whom self-determination meant freedom and in whose ra-

tionality lay the power not just to know nature but to transform it. As Karl Marx said, "The bourgeoisie, during its rule of scarce one hundred years, has created more massive and more colossal productive forces than have all preceding generations together." The scientific method stems from the sovereign individual's power to construct systems of knowledge through the empiricist methodology, a cornerstone of the scientific revolution.

Max Weber chronicled how capitalism flowed from this new conception of the individual and how the Protestant ethic provided the rationalization of life. The sovereign individual became economic man, a self-interested maximizer of utility. Ultimately, this led to Joseph Schumpeter's notion of adversarial democracy in the mid-twentieth century, a view of society as nothing more than a marketplace for self-interest.

Other societies have a different view of the individual. For example, the anthropologist Clifford Geertz found that in Balinese culture the concept of the unique individual self played only a minimal role in everyday life. The Balinese see individuals as representatives of general social categories. For them, to love or hate someone because of the state of his or her individual mind would border on the absurd.

In the West, prompted by the work of George Herbert Mead and others, the heads of sociology and anthropology viewed the self as increasingly embedded in a social order, and its relation to the social group became increasingly central. In the most influential sociological work of the 1950s, David Riesman's *The Lonely Crowd,* Riesman laments the decline of the inner-directed character type. An outer-directed man was replacing this autonomous individual: William Whyte's "company man" who lacked an internal guide, an authentic self, and instead was described by such negative characteristics as superficial, conformist, and submissive to the power of the group. Around the same time, Solomon Asch conducted his classic study on conformity in which he asked subjects to judge which of two lines drawn on a page was the longest. When other participants, who were secretly in on the experiment, unanimously chose the obviously shorter line as the longer of the two in front of the subject, that subject, too, would choose the shorter line and would literally believe it to be shorter. Mass conformity and obedience to authority marked the fall of the sovereign individual.

With this shift in perspective, the psychoanalyst Erik Erikson suggested that youth was a time of identity crisis. The development of the self became a journey involving differentiating oneself from the group. To possess a shaky sense of identity, what Erikson called "identity diffusion," signaled a failure of personality development. What most challenged construction of a firm identity was the narcissistic personality, one who constantly required approval and lacked inner conviction. The self in mass society, then, was neither totally subsumed by the collective, what Erich Fromm described as one's desire to escape the freedom of self-determination and the responsibility it brought, nor self-actualized as humanistic psychology proposed. Instead, the self seemed caught in a tension between identifying with the group and being an individual.

In trying to understand how the insights of cultural biology fit into this long debate over the sources of the self, we began to see the glimmer of a solution. All of the evidence points toward the prefrontal cortex as the center of a system that adjusts your behavior to match the contexts in which you find yourself. Building your identity requires a long interaction with a social environment, during which flexible self-representations are constructed. Twentieth-century anthropologists such as Margaret Mead pointed to this capacity and explored how different selves emerge in different cultures. But they went too far in asserting that this special capacity unfetters you from your biology. Their failure lay in not seeing how your internal guidance system is culture's equal partner in building your sense of self.

The marriage of an internal guidance system embedded in ancient diffuse neuromodulatory systems with a user's guide to life residing in a more recent and malleable cerebral cortex may have its problems. Since Plato, who in *The Republic* highlighted the mental clash between desire and reason, one type of explanation of mental conflict has centered on conflicts among different elements of the mind.

There is something compelling about the notion that human transgressions might sometimes be rooted in multiple brain systems vying for behavioral control. Brain damage often results in the inability to inhibit behaviors or to monitor one's own behavior and modify it when appropriate. The capacity to inhibit behavior stems from the prefrontal cortex, which in full bloom underlies your ability to curb your impulses for immediate gratification, project your self into the future, and make plans for retirement and other long-term interests.[32] Yet why do we often engage in behaviors contrary to our long-term interests? In the extreme, the breakdown of behavioral inhibition manifests itself in what is known as impulse control disorders. These are similar to addiction, except that in addiction the dopaminergic system is hijacked directly by a drug, whereas in impulse control disorders, it is hijacked by an external stimulus or event whose rewarding value becomes so great that it leads to literally self-destructive behavior—behavior that damages the long-term interests of the self.

Consider, for example, the consequences of the rise of legalized gambling in the United States. In 1998, Americans wagered more than $550 billion on legal gambling, a 3,200 percent increase since 1976. By some estimates, gambling, not drugs, is America's fastest-growing addiction, especially among teenagers, with the rate of pathological gambling among high school– and college-aged youth about twice that of adults. According to Howard J. Shaffer, director of the Harvard Medical School Center for Addiction Studies, "Today, there are more children experiencing adverse symptoms from gambling than from drugs." Whether or not gambling is a bona fide addiction, as it doesn't involve a substance, the American Medical Association recognizes pathological gambling as a diagnosable mental disorder.

Perhaps the lure of gambling lies in the dream of quick riches, which affects our user's guide by dangling new long-term goals in front of us. This doesn't strike us as very plausible. After all, if it's the dream of quick riches that entices us, why isn't there an Entrepreneurs Anonymous? And why would gambling have such a hold over teenagers, who may want a quick return but who typically aren't out looking for long-term financial security?

One central feature of gambling points to the involvement of our internal guidance system: The essence of gambling concerns making predictions about the future, thus involving the dopamine system.[33] Like gambling, sex can also be addictive in many respects.[34] As the next chapter will show, much of your sexual behavior is regulated by brain stem and midbrain structures, whose operation is largely unconscious. This suggests that the root of these self-destructive behaviors lies in the internal guidance system or, more precisely, in the failure of our frontal cortical systems to inhibit impulses from the internal guidance system that conflict with our longer-term goals.

T. S. Eliot used the metaphor of a sailboat for personal freedom, with reason at the helm and passions filling the sails. Your biology illustrates why keeping on course is such a struggle: not because of outdated instincts but because of how your brain has been crafted to give you flexibility to both external contexts and powerful internal motives. This also suggests why your self is not wholly a social construct unfettered from your biology.

If living with yourself can sometimes be problematic, living with others offers even more opportunities for conflicts. This raises the question of why we live together at all. Looking for an answer will take us even deeper into the human mystery and our basic needs, as we will explore in the next chapter.

Friend, 7
Lover, Citizen

The Mystery

of Life Together

Standing outside Westside Middle School, sixth-grader Emma Pittman first thought she was hearing fireworks. All around her, her fellow students were laughing, thinking the fire alarm was a prank, a welcome break from their routine. In an instant, their laughter turned to screams as they realized the popping sound was gunfire. Panicked, Emma looked up to see a rifle aimed squarely at her from the bushes. After that, her recollection gets blurry. She remembers being jolted, as her teacher, Shannon Wright, jumped in front of her and pushed her aside. She then remembers hearing the gun fire once, Shannon clutching her stomach, and then another shot, this one sending Shannon to the ground.

As she recounts the events of that day, Emma struggles

to make sense of why Shannon Wright put herself in harm's way to save her. She knows that she's alive today because of Shannon's quick action. People in Jonesboro, Arkansas, think of Shannon as a hero. That may not be much consolation to her husband and the two-year-old son she left behind.

That tragic day in Jonesboro, when two boys barely big enough to steady a rifle turned on their schoolmates, almost defies our capacity to make sense of things. As it becomes another entry on a growing list of incidents of senseless school violence, it makes you wonder: Is our social order so fragile that eleven- and thirteen-year-old boys can easily, and apparently without remorse, shatter it? Had these boys been turned into unfeeling monsters by abusive, toxic environments, or were they simply too young to know any better, too young to have yet been fully coerced by the civilizing forces of socialization? Most troubling of all, this tragedy highlights just how vulnerable life together makes us, how exposed we are to the capricious intent of others. If this is the potential cost of group life, why then do we live together?

Something else in the events of that day in Jonesboro demands explanation. Why do some people occasionally act as Shannon Wright did? Shannon Wright's sacrificing actions during the last moments of her life are surely rare, but perhaps similar motivations underlie more common acts of mutual aid. One explanation has its roots in Aristotle, who suggested that your affiliation with fellow human beings springs out of your natural sociability. Darwin would later call this natural affiliation "social instinct," which in humans underlies a moral sense that guides life just as a visual sense guides navigation. Morality would thus come naturally to us, a product of both our mental capacity for sympathy—our ability to feel the plight of others—and our social instincts. If this were so, then living together would also come naturally to us, and we would do so instinctively and spontaneously.[1]

There is another view, rooted in the Sophists of ancient Greece, expressed by thinkers ranging from political philosophers Niccolò Machiavelli, Thomas Hobbes, and John Locke to psychoanalyst Sigmund Freud. According to this view, our natural state is a war of all against all. Apparent acts of mutual aid are really no more than subtle deceptions to ingratiate yourself with others to boost your status and reputation. Cooperation is a ploy you use to gain advantage because you are at the core selfish, asocial, and filled with hostile instincts that require the coercive power of social authority to quell at least temporarily. To live together is thus to transcend our natural inclinations, requiring the regular beatings that Locke thought necessary to civilize the child.

Although this view has become popularly associated with Darwinism, the connection wasn't made by Darwin, but by his defender, T. H. Huxley. In his *Evolution and Ethics* of 1894, Huxley parted ways with Darwin over this core conception of human nature, citing the "moral indifference of nature" for why morality is not our natural state. Freud would later suggest that our nature could be repressed but never denied. For Freud, as he wrote *Civilization and Its Discontents* on the eve of Hitler's rise to power, instinct and civilization are irremediably antagonistic, locked in an eternal struggle that, abetted by modern technology, pushes us to the brink of annihilation.

Brain science is beginning to provide the first tentative answers to some of the most far-reaching questions humans have asked: Why do we live in an extended family, a community, a national society? Does our selfish and asocial nature make society a coercive force, or are we naturally social and so driven by social instincts that we spontaneously live together? Does the brain prime us for a moral life, or are our moral codes the artificial product of reason? The way in which humans have answered these questions throughout the ages has had profound consequences for how they lived, how they related to others, and how they approached problems of common concern. Today, among the most pressing issues in social science is the problem of trust, the roots of human cooperation and collective action that provide the foundation of civil society. For the first time in history, brain science can begin to contribute to these sweeping concerns, as we will explore in this chapter. We'll take you inside the brain to uncover the mechanisms underlying your social nature. We will look for clues in the neurochemistry of love and sex, the bonds between parent and child, and the biological drives that propel you to seek the company of others. We will discover that these first clues from the brain indicate that the modern image has led us astray, pointing instead to the tradition Aristotle began over 2,300 years ago, with far-reaching implications for who you think you are.

The Origins of Society

The modern answer to the mystery of life together owes its basic outline to the account that Thomas Hobbes offered in his 1651 *Leviathan*. His starting point involved a thought experiment: What would life have been like before the authority of the state?[2]

Events of Hobbes's own times colored his response to this question in two basic ways. The first was the rise of the scientific method, which drove the seventeenth-century scientific revolution.[3] Much of that method relied on taking a reductive stance toward nature, breaking the complex into simple, basic units. In physics this meant seeing objects as aggregates of indivisible atoms. In human affairs, it meant building a notion of society based on an understanding of the individual. Hobbes thus began his great political treatise with the individual, a radical and strikingly modern step. According to the Christian doctrine dominating Hobbes's day, societies were organic wholes with individuals as a part of the body of Christ. Individuals ultimately derived their identity from that larger collective vision: Each part had no shape except by relation to the social whole. Hobbes reversed all that, putting the individual before society and seeing society as nothing more than an aggregate of individuals.

The other great influence on Hobbes's view was England's civil war, which in his words had made England a world of "masterless men," with social stability muddied by factions vying for power. Thus, Hobbes colored individuals as self-interested and asocial.[4] Just as the atoms of the physics of his day were constantly in motion, so too Hobbes's individuals were filled with internal drives that kept them in constant motion. And just as atoms were constantly colliding with one another, so too under the impetus of unquenchable drives, "a perpetual and restless desire of power after power," Hobbes's individuals were constantly coming into contact. The inevitable result was conflict, leading to his vision of a war of all against all. This conception of the individual led to a great riddle for Hobbes: How can self-interested, asocial individuals filled with unquenchable drives ever peacefully coexist?

Freud wrestled with essentially the same riddle in *Civilization and Its Discontents* with one key difference: Whereas Hobbes believed humans were rational calculators of their self-interest, Freud believed that civilization stems from a primordial guilt that first arose with patricide, perhaps as a band of sons rose up to kill their father. Freud speculates that in the aftermath of that bloody act, feelings of shame so overwhelmed them that they formed laws and social institutions to prohibit such acts. Freud thus located our civilizing tendency in guilt, an emotional impulse. In contrast, for Hobbes rationality ultimately saves humans from themselves. Reason moves Hobbes's individuals, driven by the selfish desire for self-preservation, to relinquish liberty for security, ceding absolute control to a sovereign, a Leviathan, in exchange for security. The basis of life together is this social contract, in which the state exists only to safeguard the individual's self-preservation. The social contract so formed, however, is utterly artificial, motivated not by a social instinct to associate spontaneously but by deliberate calculation based on self-interest.

Hobbes's view was so appalling to his contemporaries that "Hobbism" became grounds for expulsion from political or religious service. However, key notions of Hobbes's view would become the cornerstone of Western political thought. In particular, Locke retained Hobbes's ideas of social contract as the glue of society, but attacked his vision of absolute monarchy, arguing that it simply transferred the war of all against all to one between the monarch and his subjects. Locke thus argued for a limited, constitutional government, which is in essence the modern limited, liberal state. From Locke it is a short step to the constitutional experiment of the United States, though the framers of the Constitution grappled with the two conceptions of society. Indeed, Thomas Jefferson deplored Hobbes's view of human nature, and when he replaced Locke's pursuit of property with the pursuit of happiness, it was a reference to another tradition, stemming from Aristotle, that sees us as social by nature, with happiness derived from civic life.[5]

As the colonies struggled to strike a balance between self-interest and the public good, far away in England Adam Smith was writing *The Wealth of Nations,*[6] the landmark work on the division of labor and economic productivity. Writing at the dawn of the Industrial Revolution, Smith reconceived the relationship between self-interest and social order. Whereas Hobbes believed that only absolute rule could curb self-interest, Smith saw self-interest as the basis for social order: in his famous words, "It is not from the benevolence of the butcher, the brewer, or the baker that we expect our dinner, but from their regard to their own interest." The invisible hand of the marketplace thus replaced the sovereign Leviathan, and the common interest simply flowed out of the collective pursuit of self-interest.

The Emergence of *Homo Economicus*

Adam Smith's work made the connection between our economic, social, and biological nature, a merging of ideas that would reverberate throughout the search for the foundations of social order. Economic theory may seem far removed from biological views of human nature, but the distance is diaphanous. Indeed, the influential economist John Maynard Keynes once quipped that *The Origin of Species* was "simply Ricardian economics couched in scientific language." In fact, it was after reading Thomas Malthus's *Essay on the Principle of Population,* which explored the misery and the scarcity of resources caused by overpopulation, that Darwin switched his perspective from thinking about competition among groups to competition between individuals.

Few unions of ideas more profoundly shaped the twentieth century than the merging of biological and economic man. At the level of individual, or microeconomic, behavior, this has become known as rational choice theory, whose core idea is that humans are rational maximizers of self-interest. Cast at a wider level this theory has become known as public choice theory. According to this view, for which James M. Buchanan won a Nobel Prize in 1986, voters, party leaders and other politicians, lobbyists, and bureaucrats do not act for a common good when they make public choices. Instead, they act for their own interests, whether it be maximizing their department's budget or guaranteeing their reelection.[7] With this, the foundation of social life on private self-interest was completed.

Whither the Common Good?

The rise of *Homo economicus* clashes with another conception of the origin of society. According to this idea, what separates human conduct from animal instinct is the human capacity for morality, the crux of which is human concern for others.[8] Whereas animals may be motivated by various needs, humans are supposed to act for reasons drawn from principle. Human morality is thus rooted in your duty to follow the "right" course of action, regardless of whether it is in your selfish interests to do so. Human nature thus creates a common good, with the hallmark of moral conduct being the altruistic act.

We say that people act altruistically when their action serves the interests of another person at the expense of their own. From an evolutionary perspective, someone acts altruistically when their actions enhance someone else's fitness at the expense of their own. This perspective would have you consider, for example, the soldier who goes to battle and risks life and limb because he or she believes it is the moral thing to do. The driving goals are often abstract: fulfilling an obligation to country, the defense of principles, or upholding the honor of a deity.

But if evolution is a struggle for survival in which the fittest prevail, how could such self-sacrificing behavior—indeed, how could morality itself—ever evolve?

One possible answer comes from Darwin himself. He presented his speculations in *The Descent of Man,* where he suggested that a tribe composed of members willing to "sacrifice themselves for the common good would be victorious over most other tribes; and this would be natural selection." Called group selection, the idea is that just as individuals can be pitted against one another to create selective pressures, so too may groups be pitted against one another to create group selective pressures.

Group selection looks like a way to keep the kinder side of human nature from melting in the heat of evolution. Life may still be a struggle, but if it is a battle waged among groups, then mutual aid within the group might still flourish. References to group selection were common from the time of Darwin through the 1960s, when the most forceful critique came from George Williams, then a postdoctoral student at the University of Chicago, a center of group-level evolutionary thinking. Williams was growing more and more frustrated by the loose way selection was talked about. The result of that frustration was his 1966 book, *Adaptation and Natural Selection,* which Williams hoped would bring rigor to the notion of adaptation. In addition to becoming a classic in evolutionary theory, it also effectively banished group selection from evolutionary thinking. Group selection became an evolutionary gaffe.

Following Williams's devastating critique of the idea that animals act for the good of their species, it became popular to translate apparent acts of altruism into subtle cases of self-interest.[9] This translation was greatly facilitated by the introduction of game theory into evolutionary thinking in the 1970s, an approach that became evolutionary game theory.[10]

Modern game theory stems from John von Neumann and Oscar Morgenstern's seminal 1944 work *Theory of Games and Economic Behavior,* which gained widespread adoption within economics. In game theory, two or more rational self-interested maximizers face off against one another and the problem is to find optimal strategies to win the game. The principles of game theory are applicable to evolution, since evolutionary biologists think of the game of life as a contest to get the most copies of one's genes into the next generation. Questions regarding moral behavior can be investigated within this framework by asking what would happen if a few manipulative self-seekers, known as hawks, were placed within a population of kindly altruists, the doves. Game theory demonstrated rigorously that since the hawks will use the doves for their own ends, outmanipulate them, and especially outre-produce them, over the course of generations, fewer and fewer doves will be present. In other words, altruists who act for the good of the species are an evolutionary dead end.[11]

This was yet another striking union of economic theory and evolu-tionary views pointing to the conclusion that self-interested behavior was both universal and inescapable. And it is here that the grimmest social implication was about to be reached. Consider the following story.

Hobbes's Prison

You've just been arrested. You and an accomplice are taken downtown and put into separate interrogation rooms. As you sit there alone, con-templating your fate, in walks a stone-faced detective. He tells you that if you confess that the two of you committed the crime and your ac-complice denies it, he will let you go free and give your accomplice a five-year sentence. If you both deny the crime, he says, the police have enough circumstantial evidence to put both of you away for two years. If you both confess, then you'll both get four-year sentences. The dilemma facing you, then, is whether you turn on your partner in crime.

If you both remain silent, you'd both get two years. But now you start to think, what's your accomplice going to do? The worst outcome would come by remaining silent and having your partner rat, in which case he'd go free and you'd get five years in jail, the sucker's payoff. There seems to be only one thing to do. No matter what your friend does, you'll be better off ratting on him.[12] Of course, your friend is going to do the same, so you both end up with four-year terms. But wait. If you both had remained silent, you both would have gotten just two years. Somehow, following individual self-interest leaves both you and your accomplice worse off than if you chose another solution. You both know this, but there seems no way out.

Two scientists at the Rand Corporation, Merrill Flood and Melvin Dresher, formulated this problem as a game in 1950. A colleague of theirs, Albert Tucker, helped create the story involving two prisoners facing the choice between cooperating with one another or defecting. Known as the Prisoner's Dilemma, it is among the most studied of all strategic games because it distills the essence of the paradox of group life: the clash between individual self-interest and the need to undertake collective action to solve social dilemmas.

In its simplest form, the Prisoner's Dilemma involves two persons. But many of our most pressing social dilemmas involve large groups. There are two major kinds of group dilemmas.[13] One, known as the provision of public goods, concerns the problem of free riders, who take advantage of public goods, from public TV to public highways to public schools, by using these goods without contributing to their creation or maintenance.

The other, known as a commons dilemma, has its roots as far back as Aristotle, but was made famous by Garrett Hardin's classic paper "The Tragedy of the Commons."[14] Hardin asks us to consider a group of herders who have open access to a common grazing land.[15] It is in each herder's interest to herd as many cows as possible on the land, thereby maximizing his return. But if all herders pursue this strategy, soon the resources of the commons may be depleted and all will suffer.

It's easy to find examples of this problem, from overfished waters to overharvested forests. Hardin believed the only way out of these dilemmas was to take a route eerily reminiscent of Hobbes's: acquiescing to outside authority, with the provision that this authority was not to be absolute but instead maintained by the consent of those governed; as Hardin put it, "mutual coercion mutually agreed on." Hardin's pessimistic conclusion arguably exacerbated the very problems it was intended to solve, a point we will explore in more detail in chapter 10.[16]

While most economists, political scientists, and sociologists believed that social dilemmas were intractable, the debate remained largely theoretical. The central issue that needed to be investigated was how people actually behaved in social dilemmas.

Escape from Hobbes's Prison

By the late 1970s, experimental work on social dilemmas began to accumulate, and it showed a surprising result: people cooperated at a much higher level than predicted by standard economic theory.[17] In fact, they often cooperated spontaneously, without the mandate of any external authority.

One of the most intriguing factors that derailed self-interested behavior turned out to be allowing people to communicate. Face-to-face communication substantially increases levels of cooperation.[18] Indeed, no other variable appears to have as consistent and strong an effect. It isn't just any sort of talk, though. When players are allowed to communicate through computer terminals by passing messages, the levels of cooperation are far below those of the same game played with face-to-face communication. Why should face-to-face communication matter? As political scientist Elinor Ostrom puts it, "Exchanging mutual commitment, increasing trust, creating and reinforcing norms, and developing a group identity appear to be the most important processes that make communication efficacious."[19]

This finding is an intriguing repudiation of the standard economic model of self-interest. But how does it relate to our evolutionary past and the brain structures underlying our social behavior? In an ironic twist, the answer was hinted at in George Williams's book. It would come from a most unexpected source: the selfish gene itself.

Whose Happiness, Whose Utility?

Imagine that you and a friend are watching a television special on Mother Teresa. Your friend comments that he's never seen such a selfish person. You ask why and he replies that she is only doing all that charity work because she gets enormous joy from it. She is just using those poor people to make herself feel good. This interpretation of apparently altruistic behavior is of course an old standard. If an action gives its actor a warm glow or maximizes his or her utility, then that action appears by definition self-serving. According to this argument, the only difference between the corporate raider Ivan Boesky and Mother Teresa is that Boesky is honest about his intentions.

Most people sense that something is askew with this argument. What is the alternative? Are the only genuinely selfless acts those done out of a purely neutral sense of duty, a kind of Vulcan indifference?[20] Suppose your job is to create a robot whose sole purpose is to be your personal valet. From watching science fiction movies you're aware that smart machines have a habit of eventually rising up and taking over, so you think long and hard about a way to avoid that scenario. One way would be to wire your robotic personal valet so that helping its owner was deeply rewarding to it. By making it other-directed, why would it ever rebel against you?

Now apply this question to yourself. Where do your values and preferences originate?[21] Although it is commonly asserted that you learn your preferences, many of the things you value are not simply learned, but instead reflect the properties of your brain's own value systems that we've called your internal guidance system.[22]

From the perspective of a selfish gene that wants to stick around, a good strategy is to promote behaviors that tend to pass the gene along. Consider child rearing. Why would any self-interested rational person want to do it? The costs are enormous, the hours are terrible, the opportunity costs (the amount lost on earned income) are astronomical, to say nothing of losing the ability to take off on weekends on a whim. Through the biology of parenthood, genes have made you their personal valets, making it rewarding for you to look after copies of them. But just as you wouldn't call your robotic valet self-interested, since it was other-directed to follow your goals, what grounds would there be for viewing parenting as self-interested? The upshot is that selfish genes can build other-directed, selfless people if that helps spread them into the next generation. Since genes don't have motivations—

they are entities without psychological states—it is misleading to impute selfish motivations to the parent. There's nothing duplicitous about what genes are up to. There's no need to suspect that behind apparent acts of kindness lurk dark ulterior motives. There's no need for Freudian psychoanalysis at the level of the gene.

The point here is that the selfish gene theory isn't an argument for selfish people. Indeed, it undermines it. This is a crucial insight of selfish gene theory, though its advocates took the theory too far by supposing that such genetic influences on behavior hardwire all of human behavior.[23] Our powerful learning mechanisms and the growth of our user's guide to life imply that our behavior isn't rigidly controlled by our genes, but it seems clear that our basic predispositions—sexual desire, appetite, and sociability—reflect basic facts of our biology. Luckily for us, as we explore next, our genes' incentive program is tied to some of our most cherished and deeply felt experiences, from the bonds of parenting to romantic love and friendship.

As we peer inside the brain for clues to our social nature, the biological bonds of parenting will be our starting point. What we shall see is that the notion of *Homo economicus* is not so much flawed as incomplete.

Social by Nature

Cultural biology points away from Hobbes and toward Aristotle, who believed that far from requiring you to conquer or transcend your nature, social life flows naturally out of it. This sociability, Aristotle believed, grows out of the child-parent bond, which is essential to an animal requiring intensive and prolonged child care, and spreads from there to create the bonds that underlie all your social life.

In *The Descent of Man*, Darwin, too, saw the child-parent bond as the source of natural morality and social life: "[T]he feeling of pleasure from society is probably an extension of the parental or filial affections."[24] This is an intriguing possibility. Perhaps the brain systems underlying the emotional bonds between parent and child also serve as the glue of group life and lie at the roots of the capacity for trust, which in turn underlies social capital, the network of norms forming the base of a cohesive and productive society. We suspected it is no accident that you, as a member of an exceptionally large-brained species requiring prolonged and intensive parenting, may have inherited an extraordinary capacity for trust and civic life from the heavy demands of child rearing. Indeed, perhaps these child-rearing pressures required a new form of social organization that helped shape human sociability and the capacity for trust.

From the perspective of cultural biology, the key to creating your large brain with its powerful mental abilities lies in its extended development and how this allows brain development and culture sufficient time to interact to build the social mind. However, a long childhood comes at an enormous cost by placing a tremendous burden on parents. How was this responsibility met? Consider your cousin, the chimpanzee. The male chimpanzee's contribution to child rearing ends after a brief encounter, leaving the female to care for her young alone. Because of the extended effort that child rearing places on female chimpanzees, their interbirth interval is about four to eight years, placing chimpanzees at the edge of zero population growth. In contrast, in Ache women in Paraguay, who are representative of members of a traditional society, the interbirth interval is about three years, making their reproductive prospects much more promising. Given the extra burden of human child rearing, how did we become so reproductively successful?[25]

Somewhere in our deep history a shift was made in child-rearing configurations, which likely had important consequences for our capacity for civic life. One clue to uncovering this shift lies in noticing that male chimpanzees are about 27 percent larger than the typical female chimpanzee, a sexually dimorphic difference typical of polygamous species in which males do not participate in child rearing. In modern humans, the average difference is about 17 percent, which is indicative of a pair-bonding mating system.

With this link, we can fill in some of the pieces of our past. Consider, for example, the earliest hominoids, *Australopithecus*, who emerged around four million years ago in East Africa. These creatures had a brain around the same size as their ape cousins', although the hominoids walked upright. The males were nearly a third larger than the females, weighing in at about ninety pounds versus sixty-five pounds for the females, making it likely that they had a mating system similar to that of chimpanzees.

In *Homo habilis,* this gender difference was reduced to approximately the difference found in modern humans. It is also during this time that we find an expansion of the brain. Perhaps the increased parental investment required to raise a slowly developing, large-brained child was facilitated by a shift to a pair-bonding system sometime during the last seven hundred thousand years, and perhaps this shift is related to the changes in the brain.

Although most primates do not live in families, this pattern is more common in primates than among other mammals.[26] So although the ancestors of chimpanzees followed the norm, our direct ancestors adopted family values. In one plausible scenario, the climatic forces that we saw drove the expansion of the brain increased the period of postnatal development as a consequence, which would have put new pressures on child rearing. John Allman suggests that the shift in child rearing likely involved not just a shift to pair-bonding but also the invention of the extended family.[27] As Allman puts it, "The human evolutionary success story depends on two great buffers against misfortune, large brains and extended families, with each supporting and enhancing the adaptive value of the other."[28] That is, the invention of the human extended family allowed our ancestors to avoid the low reproductive potential that large brains placed on other apes. Those individuals who could cooperate directly in helping to raise one another's young would have been rewarded with the most basic of evolutionary prizes, fecundity. These cooperative bonds would have had other benefits, including creating buffers against ecological instability, and perhaps were even strong enough to drive the emergence of human culture.

Allman adds a further intriguing idea. The age at which a primate species reaches sexual maturity correlates with brain size. The bigger the brain, the later the onset of sexual maturity. By examining the correlation across a number of primate species, he noticed that if you were just another primate, and so reached sexual maturity at the age predicted for a typical primate of your brain size, you wouldn't become sexually mature until around forty-four years of age. But obviously you, as a human, become sexually mature much earlier. Reaching sexual maturity at a relatively early stage of brain development, you might not yet be ready to care for children on your own. Allman suggests that this, too, points to the extended family. With the support of a mate, siblings, parents, and grandparents, our ancestors were able to become parents much earlier than if they had to face the world alone. In fact, as Allman notes, the onset of puberty is itself dependent on nutritional status. Thus, those groups that could ensure a better food supply, through finding more productive forms of group life, would have an even greater reproductive potential.

The Ties That Bond

Perhaps the shift to extended families was mediated by the emergence of a new degree of sociability. If so, then the ability to enter into long-term reciprocal relationships with nonkin, spontaneously seeking their company and cooperating with them, would have made new forms of social organization possible. If there is evidence for this natural sociability, it would undermine the Hobbesian view of the human animal as asocial and selfish. More than two thousand years after Aristotle conjectured that the bonds stemming from parenting may underlie social bonds, brain science is taking its first steps toward uncovering the biological roots of these bonds.

Given how central social bonds are to human life, it is surprising how little we know about the brain mechanisms underlying them. In fact, virtually every form of mental disturbance involves disruptions in social attachments. As we began to piece together the research on attachment mechanisms in the brain and pondered their implications for cultural biology, we thought Aristotle's conjecture that human social impulses are ultimately rooted in parent-child bonds was plausible, since the brain typically uses what it has rather than create something from scratch. The first hints from brain studies suggest that parent-child bonds, the pair bonds of romantic love, and the social bonds of friendship may all have a common root in the brain.[29] They all activate brain systems that use the chemicals oxytocin and arginine vasopressin (AVP); endogenous opioids such as the endorphins, which mimic the action of heroin inside the brain; and dopamine. All of these brain systems can be found in nonhuman primates, but also bear in mind that in humans these systems reside inside a brain with expanded and shifted proportions. We believe that the key to human bonds lies in the relation between these ancient bonding systems and the new capacities of your enlarged cortex.

Of Sex and Bondage

We first explored the brain basis for the most elemental form of attachment: sex. Sexual identity is deeply rooted in biology. Waves of hormones sweep across the developing brain to sculpt the male and female brains. Adam may have preceded Eve, but all other brains and bodies were first organized along the female plan, with males subsequently becoming masculinized under the influence of testosterone. Surprisingly, testosterone doesn't act on the body and the brain in a uniform way. Rather, body tissues convert testosterone into dihydrotestosterone (DHT), while the brain converts it into estrogen. So it is estrogen, long equated with female characteristics, that masculinizes the brain. In fact, females who are exposed to too much estrogen, as in the case of the daughters of women who received diethylstilbestrol (DES), an estrogenic hormone prescribed in the 1950s to prevent miscarriages, show stereotypically tomboyish behavior. Although this is surely an extremely complicated story, different actions of these hormones on the brain and body can result in female brains in male bodies and vice versa. Environmental forces, too, can help shape the biochemistry of sexual identity and behavior. For example, exposing a pregnant rat to stress can significantly alter the sexual behavior of her offspring, with many more exhibiting homosexual behavior, perhaps by altering the timing of the typical fetal spurt of testosterone, which, converted to estrogen, leads to brain masculinization.[30]

In any case, testosterone and estrogen don't act everywhere in the brain. Instead, much of your sexual makeup is rooted in areas buried deep below the cortex, especially in the hypothalamus, which is where brain researchers locate neurons that are involved in gender-specific sexual behaviors.[31]

Reptilian Love

Although the search for aphrodisiacs remains an exotic one, it is easy enough to experiment with putting other animals in the mood. If you wanted to make a horny frog, as opposed to a horny toad, you could do so by injecting the chemical vasotocin into its brain. Shortly after, your experimental frog would be trumpeting its courting sounds and copulating with any nearby female frog. Although it is unknown what triggers the release of vasotocin in the daily life of a frog, testosterone is a prime suspect. For a female turtle, on the other hand, vasotocin levels surge as she reaches the beach to lay her eggs, rise while she digs a hole for her eggs, peak as she lays her eggs, and subside as she heads back out to sea.

This ancient molecule, vasotocin, is the evolutionary precursor of two closely related molecules that play central roles in mammalian sexual behavior: oxytocin and arginine vasopressin (AVP). The hypothalamus of the male brain contains more AVP than does the hypothalamus of the female brain. Under the influence of testosterone, AVP levels rise with the onset of puberty and are linked to both sexuality and aggressiveness. When AVP is injected directly into a male rat's brain, for example, the rat begins to patrol and mark his territory obsessively, and he also becomes particularly combative. Castration or damage to these AVP brain circuits, on the other hand, results in both reduced sexual interest and reduced aggressiveness. AVP isn't related to all aspects of male sexual behavior, but it appears to underlie male sexual craving and anticipation.

Although sex for an upstanding woman in Victorian England was supposed to be a matter of duty that she hoped wouldn't last too long, brain science has overturned the notion that men and women have little in common in terms of sexual enjoyment. It is not AVP that appears to underlie female sexual desire, but rather hypothalamic oxytocin, which is present in higher levels in the female brain than in the male. Estrogen and other hormones, either injected into a female rat or secreted during normal estrus, cause a tremendous increase in the number of oxytocin receptors in the medial hypothalamus, which in turn results in a cascade of changes that alter the female rat's behavior to make her sexually receptive to males. These changes include a growth in brain cells that generates a circuit, which then sends messages down her spinal cord to sensitize what is known as the lordosis reflex. As a male initiates sexual contact, this reflex in the female's spinal cord generates her receptive mating position: the arched back and raised rump that facilitate copulation. This is just one example of the structural changes that take place inside the female in response to the neurochemistry of sex. Indeed, enormous structural changes within the female brain occur both across the menstrual cycle and with pregnancy, resulting in far-reaching behavioral changes and even affecting learning.[32]

While oxytocin is traditionally associated with female bonding, both sexually and in maternal attachment, it is not a uniquely female chemical. If we were to monitor a typical male brain during the anticipatory stages of sex, we would first see a surge of AVP. As he nears orgasm, however, his AVP levels drop almost to what they were before thoughts of sex entered his mind, while oxytocin levels in his brain begin to surge and peak during orgasm.

There's no disputing that men typically find orgasm a pleasurable experience. The explosion of interest in Viagra has created a bestseller, used as much recreationally as therapeutically. The two prime candidates thought to underlie the rewarding aspects of orgasm are dopamine and opioids. So what does this surge of oxytocin in the male brain during orgasm signify? Cells containing oxytocin send their axons to the ventral tegmental area or the VTA, one of the brain's dopaminergic sites and itself a site of opioid receptors, which we've highlighted as part of your human core, the ancient internal guidance system. Blocking the effects of opiates markedly diminishes the pleasurable sensations of orgasm. This suggests that the concurrent release of opiates and surge of oxytocin produce the euphoric sensation of orgasm. Addicts report an orgasmic rush when they use such drugs as heroin, which in the brain is quickly converted to morphine and binds rapidly to opioid receptors.

Given both the rewarding value of sex and the surge of oxytocin in the male brain during orgasm, there is a great deal of speculation that oxytocin release in male brains during orgasm facilitates the pair-bonding that distinguishes us from the great apes. It has been harder to understand the role of the female orgasm, since it is less predictable than the male orgasm, and since it doesn't have the obvious reproductive role that a male orgasm does. Some have speculated that the female orgasm is simply a by-product of the male orgasm, while others link it to reproduction. According to one argument, sperm retention is slightly higher when a female has an orgasm.[33] Even if this is true, it's likely that a direct link to reproduction will only be part of the story. Although reproduction is the bottom line, sex plays a much larger part in human life than merely as a way to reproduce, as indeed it does in our bonobo cousins, mentioned in chapter 4. To see how it does this, we need to consider the issue of sexual receptivity in women.

As we noted in chapter 4, one of the key differences between chimpanzees and bonobos is female sexual receptivity. A female chimpanzee is sexually receptive only during estrus, which is signaled to the group by genital swelling. In contrast, the female bonobo is sexually receptive throughout much of her cycle, or more than half of her adult life compared with only 5 percent for the female chimpanzee. This extended receptivity in the bonobo translates into a great deal of what seems like casual and often same-gender sex, much of which is aimed at conflict resolution and maintaining social cohesion. In humans, women are also sexually receptive throughout their cycle, but with a quite different purpose: In humans, extended sexual receptivity is thought to facilitate pair-bonding. By being sexually receptive through-out her cycle, a female can maintain her bonds with a male through regular sexual activity, whereas a female who is sexually receptive only for a small fraction of the time will be at a disadvantage in keeping her mate's interest.

What might be the role of the female orgasm in all this? Although female orgasm was once thought to be uniquely human, a variety of studies now suggest that various other primates appear to experience it as well.[34] Among those species studied, only human females engage in pair-bonding, however, suggesting that orgasms might play a unique role in human females. In other species, female orgasm is believed to be involved in selective mate choice; that is, it may help direct females toward males with whom sex is more likely to result in orgasm. In hu-mans, the intense pleasure of orgasm as a psychological reward, com-bined with extended sexual receptivity, hints that orgasm enhances pair-bonding, though the issue remains rather mysterious scientifically. Indeed, there is a release of oxytocin in a woman's brain during female orgasm, just as there is in a male's brain, and levels of oxytocin remain elevated for some time after orgasm.[35] In this light, it is interesting to speculate that the relatively lower predictability of female orgasm may underlie selective bonding with those who can elicit it. Indeed, the number of submissions to advice columns from sexually unsatisfied women reveals just how disruptive the lack of satisfying sex can be to pair-bonding.

At first glance, it might seem demeaning to think that romantic love has its roots in these chemical processes inside your head. Surely, there is more to love than ancient neuromodulators. In humans, both thought and imagination color romantic life. What we were trying to understand, though, was the basis for impulses that drive you toward others and the emotional force these impulses possess. Before thought or rational deliberation enters into your relationships with others, it seems that deeply rooted impulses first act by creating the desire to be in the company of others.

Our intuitions were strengthened when we learned of an elegant experiment that utilized the ability to create transgenic animals, that is, to insert genes from one species into another. Thomas Insel and his colleagues at Emory University took a gene that codes for a type of AVP receptor from a monogamous species of vole and inserted it into mice. The mice grew to have a similar pattern of AVP receptors to the vole's. Since the vole's affiliative behavior can be increased by administering AVP, Insel and his colleagues examined what would happen if AVP was administered to the transgenic mouse. Amazingly, the transgenic mouse showed an increase in affiliative behavior, just as the vole did.

We believe that these results are a striking example of how a single gene can alter complex behaviors, perhaps by making circuit changes that alter behavior patterns that a species finds rewarding. But can we say that the unique ways these chemicals work in your brain underlie romantic love? Cross-cultural and anthropological studies confirm that the idea of romantic love is found throughout history and across diverse cultures. But what is love, anyway? At its core is the notion of exclusive commitment, of channeling one's sexual, emotional, and economic resources to another. Although brain science will never compete with the silky lyrics of love ballads, seeing how some of the brain's most powerful systems may lie at the roots of love, Robert Palmer may have a point in his crooning about being addicted to love.

When in love, do you literally become addicted to another person? Although dopamine systems are central in drug addiction, dopamine itself doesn't appear to mediate the experience of pleasure. One prime candidate for this pleasure experience is the brain's endogenous opioid system, the endorphins, enkephalins, and dynorphins. The opioid system, a system that evolved to modulate and alleviate the sensation of pain, can produce the powerful analgesic effects of morphine. At some time in your life you've likely felt the pain of separation from a loved one, and perhaps wondered what within you could cause such agonizing feelings of loss.

Cambridge University's Barry Keverne is investigating attachment in female talapoin monkeys, which develop what we would recognize as deep friendships with one another. Normally, they spend a third of their day grooming and intertwining their long tails. When two talapoin friends are separated for a long time and then reunited, they immediately begin to groom one another enthusiastically, causing their endorphin levels to double. When such separated monkeys are given a dose of morphine just before their reunion, however, they don't groom each other. Keverne suggests that since morphine has already boosted their endorphin levels, grooming no longer has its normally rewarding value.

In humans, morphine is a powerful alleviator of grief and loneliness. All this raises the possibility that those who cannot derive satisfaction from natural social relationships, with their endorphin-releasing effects, may turn to drugs of addiction, such as heroin or opium. It may also be why some addicts lose interest in social ties and so let their social life fall apart.

We're not suggesting that AVP, oxytocin, and endogenous opioids are all there is to say about human love. But we do think that humans feel these emotions so deeply precisely because they are rooted in chemicals that have immensely powerful effects within the brain. Your brain primes you to love, makes your intimate encounters among the most deeply felt of your experiences, and even makes you physically dependent on another person. Indeed, although love is often contrasted with hate, grief would seem the more appropriate opposite. Given what brain science is revealing about loving feelings, it is no wonder that we talk about the pain of personal loss, a feeling that has more in common with physical pain than we might have expected.

As we pondered these powerful brain systems, we realized that they had the necessary properties to underlie human sociability, and perhaps even the affiliative impulse that can build the trust needed for a civil society. They certainly create enormously powerful social impulses, and the experiments of Thomas Insel and his colleagues demonstrate that evolution could have reconfigured these systems relatively easily to generate novel social arrangements.

Nurturing Bonds

If the attachments between parent and child shared common mechanisms with those of romantic love, then our hypothesis would require only one more step: from familial attachment to friendship.

When we talk about parenting practices, it's important to bear in mind that they are a complex blend of social, biological, and economic influences. Looking across cultures and history, we find an enormous variety of child-rearing practices. Although Margaret Mead and other cultural anthropologists took this variety as evidence for the social construction of parenting, none of it really means that your biological heritage has been marginalized. Rather, from the perspective of cultural biology, the variety of parenting we find is perfectly fitting for the human animal. Since evolutionary success derives from the brain's capacity to shape circumstances with cultural institutions and practices, one would expect parenting—and the conception of childhood itself—to be a rich combination of biological priming and cultural problem solving. Indeed, human mating strategies themselves, though showing many commonalties across cultures, are not locked in place by biology, but are influenced by social and economic conditions.

Consider, for example, the well-known finding from evolutionary psychology that each sex has its own mating strategy, with men typically seeking young women with high reproductive potential and women typically seeking men with resources to invest in offspring. Why should these preferences be considered the result of an evolutionary module when generic problem solving could result in similar strategies? A few years ago, in the midst of welfare reform, many politicians blamed the woes of the welfare system on its skewed economic incentives, which rewarded single women for having children, allegedly leading to the proliferation of dependent "welfare moms." On the other hand, increasing numbers of women with their own substantial resources are now deciding to bypass traditional relationships to have children on their own, reflecting the human capacity to shift mating strategies in the light of differing economic conditions.

Still, there's little doubt that the desire to have children reflects a basic biological need. In this postindustrial society, in which the total cost of raising a child is roughly $1,500,000, the overwhelming majority of people continue to have children, some investing many thousands of dollars in fertility treatments.[36]

Cultural values can overwhelm the biological desire for offspring, as is tragically evident inside some of China's orphanages. China's strict birth restrictions have put baby boys at a premium. Girls, who do not carry on the family name or traditionally provide for their parents in old age, are being abandoned in alarming numbers. Overrun, the orphanages have become an institutionalized form of infanticide. Babies who are not quickly adopted are moved into the toddlers' room, a grim place typically furnished with row upon row of cribs. There, care and contact are minimal, with only three or four women in charge of almost fifty children. At feeding time, bottles are placed in the cribs for the children to use without assistance and are collected according to schedule, used or not.

Beyond the toddlers' rooms, an even grimmer fate awaits. These areas are known simply as the dying rooms. Tucked behind the main buildings of Chinese orphanages, they are hidden out of sight from the casual visitor. Inside the horror isn't just the suffering, it is the painful silence of children withdrawn into their shells, their bodies stiff and motionless. Abandoned by their parents and passed over for adoption, they spend their last days slowly starving not only for food but for human contact, both deprivations equally deadly. When human contact is denied and the desire for attachment goes unanswered, this is the result: children who literally shrivel up, first into themselves and then into nothingness. Whereas it was once thought that children bond to their parents only through experiencing the rewarding comforts parents provide, these tragic cases illustrate that the brain contains both a social reward system that propels you to seek the company of others and a distress system that makes separation psychologically painful and potentially life-threatening.

What is so disturbing about the dying rooms is that the abuse is passive. These children aren't subjected to physical abuse in the normal sense. They are simply neglected, denied human contact. And although the problem of the Chinese orphanages is shocking because it is a country's institutional policy, too often similar horrific damage is inflicted by patterns of neglect in the home. The results can be found in any mental health facility of any hospital in the United States.

The case of "Rachel" is typical.[37] Nine years old, Rachel, neglected as an infant, had been bounced from foster home to foster home, each unable to deal with her uncontrollable rage. Finally, flailing and kicking, she was admitted to a hospital. Put into a quiet environment, she sat with arms around her knees, rocking back and forth, staring vacantly. She was diagnosed as suffering from attention deficit/hyperactive disorder coupled with a conduct disorder. A month of behavioral therapy was of little help—she still lashed out at anyone within striking distance for no reason. Finally, drugs to stimulate her serotonin system were administered, which eventually quieted her.

Decades of animal studies have revealed the numbing effects of neglect, disturbingly like Rachel's, in other primate species. When monkeys aren't allowed to form bonds during the first six months of life (equal to about the first two years of human life), they show the same rocking behaviors of neglected human children. And they show signs of severe social disturbances—withdrawal, impulsiveness, and unprovoked aggression, as well as learning disabilities.

We've all seen just how helpless newborns are, literally unable to control their limbs or hold up their heads. Yet they are born with eyes open, from the moment of birth ready to gaze intently at faces, engrossing the caregiver and child in each other. Within minutes of birth, the newborn is able to imitate facial expressions. By just six weeks after birth, the newborn returns the caregiver's smile—the first sign of a positive social emotion, called positive affect by psychologists. For the next several months caregiver and child spend their time locked in interpersonal engagement. As we chronicled in chapter 6, at around ten months of age an infant's world expands to engage caregivers with objects of mutual interest, social interactions that provide the scaffolding for building the human mind.

From the perspective of cultural biology, engaging in this social web of relationships, first between caregiver and child, and then expanding to include the school, the community, and, ultimately, society at large, is crucial. Your brain made the gamble that it would develop in this rich nexus of social relationships, and so it is essential that it have mechanisms for attachment and social engagement. This is the job of your internal guidance system, which utilizes many of the chemicals that underlie other forms of social engagement, including AVP, oxytocin, endogenous opioids, and dopamine. For the infant, then, these fuel the motivation for social inclusion, attachment, and affiliation that drive the unfolding of a person.

If the child is drawn into the social world from within, what in the biology of parenthood draws the parent to the child? One early clue came from the surprising finding that transfusing blood from a rat that had just given birth to a virgin rat elicited maternal behavior in the virgin. These results suggest that the hormones estrogen, progesterone, and prolactin, all coursing through the bloodstream and rising and falling at various stages of pregnancy and motherhood, underlie key components of maternal behavior. In particular, during the final stage of pregnancy, estrogen surges cause a huge increase in oxytocin receptors and even in the numbers of neurons in the hypothalamus that make oxytocin. Recall that oxytocin is involved in sexual behavior, and perhaps pair-bonding in humans. It is intriguing that this chemical is also involved in maternal behavior. In that context, it is best known for its role in promoting uterine contractions and milk letdown. Indeed, pitocin, the synthetic form of oxytocin, is routinely administered to women in labor to induce contractions. In addition to these physiological functions, oxytocin is also involved in inducing maternal mood and attachment. Indeed, injections of oxytocin directly into the brain structures that sit just above areas underlying sexuality can cause maternal behavior, while damage to oxytocin-producing sites can stunt maternal behavior.

As we explored the role of oxytocin in maternal behavior, we found more evidence to suggest that the diverse forms of human bonds may spring from a common neurochemistry. For example, there is a suggestive link between oxytocin and the ventral tegmental area (VTA), a primary component of your internal guidance system, in maternal behavior.

The projections of dopaminergic neurons from the VTA to many parts of the brain strongly influence goal-seeking behaviors, directing you toward valued objects in the world, and they are also involved in maternal behavior toward the infant. In addition, the strong emotional experiences underlying parental attachment appear to be mediated by the concurrent release of oxytocin and opioids, another powerful indication of the common mechanisms underlying human social behavior.

Often biology reveals a beautiful synergy that inspires admiration of its harmony. As we looked into the chemistry of human bonds, we found a striking example of this harmony. As an infant feeds at her mother's breast, the suckling action stimulates the release of oxytocin into the mother's bloodstream. As we have mentioned, the most obvious physiological effect of this oxytocin release is a milk-letdown reflex, but oxytocin is also passed through the mother's milk to the child. The result is that both mother and child receive a simultaneous surge of oxytocin during nursing, which may facilitate their mutual attachment.

Experience and Human Bonds

As potentially powerful as these attachment mechanisms are, the existence of orphanages like those in China raises another basic question: Why is it that under some conditions parents will give up their infants, presumably with full realization of the devastating consequences to follow? There has been a tremendous amount of controversy over the issue of human infanticide. Its existence requires us to confront the question of why parental behavior hasn't been built in to make it more automatic.

Part of the answer, we believe, requires abandoning the simplistic notion of instinctual behavior as a hardwired reflex. Many human maternal behaviors are not instincts in this sense; instead, brain mechanisms have been identified that tend to orient and motivate the mother toward the infant, priming her for the acquisition of parental skills. Some of this learning is spurred by making interactions with the infant rewarding,[38] but the shift from conditioned control to cognitive control over behavior allows a mother to integrate a wider range of information, such as medical advice and experience passed from mother to daughter. Thus, the trend in primate parenting that culminates in human parental behavior is a move away from a strict hormonal regulation of behavior toward one of cognitive control.

The hallmark of competent parenting is maternal experience, which begins with adolescent females learning by example. If these skills are open to cultural learning, humans can adjust parenting practices in response to changing social and ecological conditions. Of course, this flexibility comes at a cost: In the absence of the requisite experience, incompetent parenting or strong cultural pressures greatly increase the risks of tragic outcomes, like the Chinese orphanages and equally horrific examples of child abuse that have been documented in our society.

We found intriguing evidence for this point of view in oxytocin's involvement with maternal behavior. Blocking oxytocin interrupts the acquisition of maternal behavior. Once those maternal bonds have been established, however, blocking oxytocin doesn't stunt maternal behavior. This suggests that oxytocin is essential only to initiate maternal behavior and to support the learning phase. Once maternal behavior has been learned, other mechanisms maintain the behavior. This new view can be tested with experiments designed to assess how internal mechanisms and cultural systems interact in guiding maternal behavior.

When Friends Get Inside Your Head

The organization of the hormonal and neural systems underlying human social bonds vindicates Aristotle's conjecture that human sociability flows out of the bonds of child rearing.[39] The missing link that we needed to determine is whether the systems underlying parent-child attachment might also underlie other forms of human affiliation, such as friendship. Many other species have evolved ways to recognize their own kin and thus restrict their confederations. Do such discriminative mechanisms operate in us? Consider the following scenario.

Marlene was overcome with grief at Max's funeral. He had always been there for her, a constant in her life that she now had to face without him. As he was gently lowered into his final resting place, she turned to walk away, but not before one final, tearful look at where Max, her beloved schnauzer, would lie for eternity.

Most people feel tremendous grief over the death of a pet. According to one study, 18 percent of owners were so stricken that they were unable to carry out their daily routine.[40] From an evolutionary perspective, though, pets are a mystery. If you care for your children because evolution favors those who help propagate their genes, why waste so much time and resources caring for your pet, since they share so few of your genes?

There are around 112 million cats and dogs in the United States. In New York and other large cities, where life can be lonely and alienating, there's approximately one cat or dog for every two people. People pay a staggering $6 billion or so annually on vet bills alone. In fact, in a country in which health insurance is elusive for tens of millions of people, it is amazing that such companies as Veterinary Pet Insurance and PetsHealth Insurance offer pet insurance.[41] Pet owners also increasingly turn to holistic vets, nutritionists, aromatherapists, and acupuncturists for Fido, not to mention pet beauty salons, gourmet food, and designer accessories. In one survey of dog owners, 48 percent defined their dog as a member of the family, 67 percent displayed a photograph of their dog, 73 percent let the dog sleep in their bedroom, and 40 percent celebrated its birthday.[42] This is an extraordinary investment of resources. Most perplexing of all is the willingness of some pet owners to risk their lives rescuing pets from fires and other disasters. If altruism toward other humans is hard to explain, what could ever explain altruism toward pets?

Among competing explanations, the most plausible is that pets are "parasites" who take advantage of human parenting attachment mechanisms to get you to care for their needs.[43] Although calling your beloved pet a parasite doesn't exactly sound like a compliment, no bad intentions on their part are implied. In fact, dogs are intentionally bred to possess childlike features. Watch almost any dog and owner interact and you'll see striking similarities to the parent-child bond, from the use of "parentese"—the high-pitched, exaggerated way we talk to children—to the way we refer to Fido as our "baby," "boy," "cutey," and the like.

The fact that you so readily form attachments with your pet is strong evidence that human social attachment mechanisms do not finely discriminate between kin and nonkin. If pets can substitute for children, then the same mechanisms can be extended to friends, some of whom are as loyal as Lassie. Indeed, there are good reasons for why we do not instinctually discriminate between kin and nonkin. For one, cohorts in tribes were likely to be genetically related. Moreover, the ability to form such attachments would have primed our ancestors for social cooperative ventures that enhanced their own survival. We know that for apes the ability to form coalitions is more important than being physically dominating, so human prosociality would have large advantages. Perhaps most important, a shift from instinctual control of attachment to cognitive control means that the maintenance of friendships would depend on experience and cognitive evaluation. Although permissive attachment mechanisms might sometimes get you into sticky friendships, your ability to evaluate those friendships would let you successfully navigate them as well as select friends likely to be supportive.

Looking for the roots of friendship in the chemical nexus of AVP, oxytocin, dopamine, and opioids made us realize that Aristotle's conception of human nature was closer to the biological truth than Hobbes's view that dominates modern times. Far from making you an asocial animal, the chemicals inside your head propel you to stay with your family, make the company of others rewarding, and foster the bonds that underlie trust. There was something else, though, that made us wonder whether human social bonds were important in unsuspected ways that could even affect our health.

Why Your Friends May Save Your Life

A conversation with Caltech's John Allman sparked the question of why women in the United States on average outlive men. Although this is sometimes thought to be a recent phenomenon, rooted in improvements in medicine that make childbearing less risky, Swedish mortality records going back to 1780, long before modern medicine, show that women lived on average 5 to 8 percent longer than did men. In a study of ten primate species, including humans, apes, and various Old World and New World monkeys, Allman and his colleagues showed that the parent that cares for the offspring typically lives longer than the mate, regardless of gender.[44] The mother is not always the primary caregiver. The male Titi monkey of South America, for example, who takes care of the baby after the mother has given birth, outlives his mate by 20 percent. In contrast, female chimpanzees and orangutans, who are the sole caregivers, live significantly longer than do their male counterparts. In captivity, female chimpanzees live on average 42 percent longer than do males. In the wild, three times as many females than males are found in groups, reflecting their greater survival rates. This pattern is reversed in such species as the Siamang apes, who pair-bond. Siamang males are unique among apes in the intensity of their bonding with their offspring, whom they carry from the second year on. Male siamang apes live 9 percent longer than their female counterparts. Male gorillas play with their young and protect them, a secondary role that is reflected in a smaller gender difference in mortality. Similarly, humans show a less pronounced difference than chimpanzees and orangutans, reflecting a significant male role in providing for offspring.

Allman believes that this gender-specific enhanced survival is the result of a complex blend of reduced risk taking and reduced vulnerability to stress, which may be mediated by the hormonal effects of parenting. Indeed, there is evidence that taking care of an elderly parent or even a pet may promote greater longevity, suggesting that caring for others is beneficial to one's health, although the economic conditions of today's society can also make this stressful on families.

In a hypersocial species like humans, do those who enjoy a cohesive and rich social life tend to outlive those who are lonely, alienated, or socially detached? Consider the results of a nine-year study involving almost seven thousand adults in Alameda County, California, that culminated in 1979. Epidemiologists Lisa Berkman and S. Leonard Syme reported that people with few social ties were two to three times more likely to die of all causes than were those with more extensive contacts.[45] This relationship persisted even after controlling for such characteristics as age and health practices, including cigarette smoking, drinking, exercise, and the use of medical services. The basic findings of the Alameda County study have since been confirmed in more than a half dozen epidemiological studies in different communities.

In recent years, the nature and quality of social ties have come to be known as social capital, defined as the social networks, norms, and trust that enable groups of individuals to cooperate in pursuing shared objectives. Political scientists such as Harvard's Robert Putnam have suggested that social capital, like physical and human capital, is a major factor in economic development.

Ichiro Kawachi, Bruce Kennedy, and Kimberly Lochner of the Harvard School of Public Health examined a further link: that between social capital and health.[46] They looked at the relationship between the level of civic trust and the age-adjusted rate of death from all causes for the thirty-nine states for which data were available in the National Opinion Research Center's General Social Surveys. The lower the trust, the higher the average mortality rate. They also found that mortality rates are linked to membership in voluntary associations—the lower the membership, the higher the mortality rates. When they examined the relationship between social capital and people's sense of well-being, they found a striking relationship between the two. (In chapter 10 we'll take a closer look at social capital and its links to the biology of human happiness.) Since the early days of life in America, Americans have traditionally been avid joiners of recreational clubs, political parties, and civic organizations, but there is evidence to suggest that recently more and more Americans are bowling alone, which, if confirmed, could have serious consequences for health and social cohesiveness.[47]

Although a great deal of work remains to untangle the relationship between friendship and mortality, we do know that oxytocin exerts potent physiological antistress effects.[48] In experiments on rats, a five-day period of oxytocin injections decreased blood pressure and lowered levels of cortisol, the body's main stress hormone. There is also evidence that oxytocin levels increase in response to antidepressants with particular efficacy in disorders of social relationships.[49] In contrast, prolonged exposure to high levels of cortisol can put you at risk for decreased bone mass, depression, and brain degeneration.

These studies strongly suggest that human sociability stems from ancient brain systems that not only underlie our sexual bonds, child rearing, and friendship but also play an important role in our physical health. They also powerfully illustrate the shift from hormonal control of behavior to a more complex form of behavioral control in which ancient systems do not simply underlie a behavior, but help prime us to learn rewarding behaviors for specific domains.

Attachment and the Symbolic Mind

One of the greatest mysteries left to solve in the human puzzle is how primitive attachment systems interact with cortical systems to underlie a moral sense and systems of ethics. One way to think about this interaction is to view the neurochemistry of human attachment as an emotional fire that drives you to engage in the social world. As these attachment systems drive social development, cortical areas add new layers of complexity to your social interactions. In particular, prefrontal areas add layers of cognitive sophistication that include our self-awareness, awareness of others as persons, long-term planning, and the ability to rapidly shift behavior in the light of changing social contexts to create a moral sense that may only be fully developed in humans. Without the deep, biological impulses to engage the world, however, these higher capacities would never take shape.

As our prosocial impulses drive the construction of more complex layers of cognitive control, our capacity to use a theory of mind emerges, with perspective-taking ability that underlies the experience of empathy: our capacity to look at another human plight and feel what that person is feeling.[50] What combination of attachment mechanisms and empathetic feeling propelled Shannon Wright to sacrifice her life to protect Emma Pittman that day in Jonesboro? We can only guess at the details, but the powerful effects are plain to see. An action such as hers doesn't seem to be the long-deliberated result of calculated self-interest, but instead appears to spring from biological feelings to care for and protect the young and to feel their suffering so strongly that one's first reaction is simply to try to alleviate it.

Our capacity for self-awareness also appears to underlie our ability to create a moral order, with norms that regulate our long-term behavior in part through defining symbolic roles for us in the family and the community. As the case of M. L. demonstrated, the loss of self-awareness impairs this capacity to navigate a complex social world. However, as we considered in the previous chapter, the stunningly flexible behavior underlying human social competence suggests that the idea of a single, stable self is illusory. Instead, each of us has a repertoire of selves upon which to draw as we pass through the human drama. Can these selves be arranged along a continuum from your most private self to a public or collective self? If so, what special role would your collective self play in your social life?

It seemed to us that the distinction between yourself and others begins to fade as the identity of the crowd and who you are merges into one collective identity, a common set of symbols shared with others. Almost everyone can recall some moment in their life when they felt this collective self envelop them, triggered at such diverse events as a religious ceremony, a rock concert, or a political rally.

By creating collective identities, humans can define groups more diverse than those based on kin, such as citizenship. This is not to say that you're as willing to take a personal risk for the benefit of a fellow citizen you've never met as you would be for a close friend or kin.[51] It is striking, however, that we use the language of kin to define such groups—for example, brotherhoods—and view the nation as a family, for metaphors often define the values we live by.[52]

Among the twentieth century's darkest lessons were those involving political systems that attempted to put the collective self before the private self. Not only does Adam Smith's belief in the motivational primacy of the private individual appear secure,[53] but the primacy of the private self appears indispensable for the securing of individual rights. Yet there is little denying that we are a deeply groupish species. Of course, from the great animal herds thundering across the savanna to the intricate workings of the beehive, other species are also groupish. Unlike them, however, we don't simply belong to groups. We *identify* with them. Recall the surprising finding from experimental work on social dilemmas: Groups that are allowed face-to-face communication show higher levels of cooperation than groups who never meet. Robyn Dawes and his colleagues believe that these high levels of group cooperation are the result of human sociality. In their words, "Ease in forming group identity could be of individual benefit. It is not just the successful group that prevails, but the individuals who have a propensity to form such groups."[54]

Biological attachment systems, a theory of mind, a rich repertoire of selves. These are the components that underlie your capacity to live in a moral order and to build moral codes. But we realized that there is one more essential component to our social order.

We've argued that humans are biologically primed to acquire behaviors through cultural learning. For this to be effective, evolution had to build strong biological tendencies to internalize cultural codes of conduct, fueled by the neurochemistry of attachment.

This predisposition is called conformity, the propensity to be socialized. Although in this age of the individual, conformity is considered a character flaw, there are actually good evolutionary reasons for it. Whereas other species pass on acquired skills through processes of imitative apprenticeship, human culture is a unique method for transmitting and sharing information. It allows us to communicate our ideas to you using symbols—in this case, ink on the page. Imagine if we had to stop by your house, cut open a brain on your kitchen table, and silently point to various structures. It is much more powerful—and less messy—just to tell you about it. Human culture literally opens the universe to human inquiry by allowing us to discuss things that no amount of pointing could ever transmit. We can talk about quarks, black holes, and things and places you could never experience directly—even things that don't exist.

The implications of this sort of transmission of knowledge are astounding. Imagine, for example, that some distant ancestor had discovered a wonderful fishing hole. He could let others know about it by taking them there. But this direct knowledge is easily lost. If, however, he could tell them that it is a long day's walk along the river down from the mountain that looks like a horse across from the great cliff, then this piece of knowledge can be passed from member to member without any direct experience of the fishing hole itself. And so, perhaps as the years pass, someone will tell a story he heard on his grandfather's knee of a great fishing hole, and how to craft a fishing spear and how to use it.

Those who are adept at symbolic learning have a huge advantage over those who aren't. Symbolic learning involves taking other people's word for it, something you do every day with astounding readiness. Recently, an e-mail arrived warning of a lethal e-mail virus. If you opened the infected e-mail, the message continued, this malicious virus would reformat your hard drive, wiping out your entire disk. Hurry, the message said, forward this warning to everyone you know. There was no such virus with such malicious intent. It was a hoax, perpetrated by someone who is implicitly relying on people's capacity to believe what they read without seeking independent verification.

The capacity to believe what you are told is especially acute in children. Tooth fairies slipping money under pillows, fat men in red suits sliding down the chimney, bunnies delivering candy and eggs—the list seems endless. But as the e-mail virus hoax shows, children aren't the only ones who can be easily socialized. The point is that the tendency to be socially receptive to information says nothing about the content of what you absorb. There is always the possibility of being manipu-

lated by others for their own ends. Despite the image of the rugged in-
dividualist, group socialization is a tremendously powerful force. Conformity permeates who we are, from worrying about the width of
neckties and the length of hemlines, to whether "our" team will win
the championship. Indeed, there is perhaps no more ominous a sight
than a stadium of football fans doing the "wave."

It would be premature to say that we know the seat of human beliefs, but a good conjecture would locate it in the intersection between
the powerful goal-seeking propensity rooted in your dopaminergic system—underlying the natural curiosity of a child who touches a hot
stove just to see what will happen—and your rich social cognitive life,
mediated by the prefrontal cortex. Together, these systems compel you
to pursue experiences and extract coherence from them, imparting a
primordial need to live in a world that makes sense to you. Perhaps the
paranoia of the delusional schizophrenic differs only in degree from
your normal everyday attempts to make sense of your life experiences.
And, like the schizophrenic who sees patterns and motives where
there are only random events, how many of your own explanations are
little more than confabulations to satisfy your need for a world with
known causes?

Your capacity to believe and your need for a coherent world make
you a member of the only species for whom socially shared ideas possess powerful rewarding value. For us humans, ideas do more than
communicate what is. They also communicate what should be, the
normative dimension of human conduct. If these normative ideas take
hold within your brain, and so become those things you value—and
they must, since you call them values, after all—then the question becomes, How far can they motivate you? In some ways, ideas can be like
drugs, co-opting ancient structures in the mind and hijacking their
power to reward and motivate. If we're right about this, then the mere
possession of an idea can override the deepest evolutionary precept as
surely as addictive drugs can. Indeed, the surest way to be condemned
is to be critical of another's belief system, so powerful a hold do ideas
have on who you think you are.

Does an adaptive story tell us everything we can know about a martyr who dies gladly, swept up in the rapture of beliefs, like Joan of Arc? And how far will you go to find a world that makes sense to you and an identity that eases loneliness and alienation? The answer is: very far indeed.

Conformity, Cults, and the Biology of Loneliness

Looking out over the southern California hills from Terry's house, you can make out the red-tile roof among the estates and orange tree orchards. Almost lost in hillsides reminiscent of Tuscany, the house would have remained unknown were it not for what police discovered when they answered an anonymous tip on March 26, 1997. Inside, they found thirty-nine bodies lying tidily on bunk beds under diamond-shaped purple shrouds. Adorning the shoulder of their black garments was an arm patch reading HEAVEN'S GATE AWAY TEAM. The suitcases at their feet hinted of a trip, all containing identification, spiral notebooks, quarters, a five-dollar bill, and lip balm. Police would learn that all thirty-nine had peacefully downed phenobarbital-laced pudding, chased with vodka.

Waco. Jonestown. Heaven's Gate. These are just a few of the names that have become synonymous with how far socialization can affect human behavior. Though you would like to suppose that the social forces working on you are totally different, the differences are of degree. What beyond degree is the difference between a cult's indoctrinating rituals and those of other, more socially acceptable groups—from fraternities to the military to the hundred thousand or so human religions that humanity has invented? And though you may believe that people are drawn to cults because they suffer from some borderline personality disorder, this, too, is a myth. In most cases, life events—the death of a relative or loneliness—make an otherwise resilient person vulnerable for recruitment. Recruitment is the right word, as the overwhelming majority of these people are recruited into cults rather than having sought them out.

Why did the Heaven's Gate Away Team so cheerfully end their lives? As investigators pieced together their bizarre world from the notes they left behind and from troves of *Star Trek* and *X-Files* videos and Internet postings, what emerged was a strange combination of Gnostic religious fanaticism and New Age lore. Imprisoned by their bodies, which they called "vehicles," they saw their actions as the ultimate transcending of the flesh. Indeed, six members of the cult had already denied their bodies through surgical castration. They were heading for a rendezvous with a spacecraft that was following in the slipstream of Comet Hale-Bopp. From Level Above Human, it would take them to a new world. No wonder they seemed so eager for the trip.

Although their departure from this earth was a peaceful one, the same couldn't be said for the largest mass suicide in modern history. Why did a stunning 912 people at Jonestown commit suicide in 1978? As one psychologist commented after interviewing Jones's followers, "They described their experiences as finding an unexpected sense of purpose, as though they were becoming a part of something extraordinarily significant that seemed to carry them beyond their feelings of isolation and toward an expanded sense of reality and the meaning of life. Belonging to the group gradually becomes more important than anything else."[55]

Thomas Hobbes was right when he said that humans contain unquenchable impulses. But among the impulses that he mentioned, Hobbes did not include the unquenchable impulse for a meaningful existence and a valued place in a social order. Loneliness brings vulnerability and the need to identify with something larger than your private existence. For the people who followed Jim Jones into the remote jungle of Guyana, absolute allegiance was demanded in return for Jones's promise of a new life on another planet. In case his followers were tempted to forget it, the message was drilled into them by ritual abuse involving public rape and humiliation, beatings, electrical shocks, and, as cult expert Keith Harary put it, "the fear of returning to an old self associated almost exclusively with feelings of loneliness and a lack of meaning."[56] For the people of Jonestown, listening to God incarnate in Jim Jones, seeking to please him and escape his wrath, was enough to die for—and to kill for. It is sometimes forgotten that there were 272 children at Jonestown. Many innocently drank the cyanide-laced concoction. But it was forced on others. There is hardly a more horrible image to conjure up than that of a small boy being held down, arms flailing and legs kicking, as his parents force poison down his throat over his screams of protest.

In trying to make sense of these bizarre and horrifying scenes, an evolutionary story can only go so far. We have seen some reasons why conformity and social needs drive much of what you do. Yet it is ultimately the infective power of ideas that gives final shape to your values and actions. None of this is to say that you have magically unfettered yourself from biology, but that sometimes your biology can unfetter you from reality itself.

The Killer Within 8

From the Solitary Killer to the Killing Crowd

The images flashing on the TV screen weren't the grainy black-and-white footage of refugees caught in the horror of Nazi Germany. These images were in color, and the men, women, and children—some herded onto flatbed trucks while others walked in a grim procession—looked unmistakably modern. The atrocities they spoke of seemed out of place against the background of modern buildings: accounts of soldiers picking out fathers, sons, and brothers and leading them off; of the gunfire that shortly followed.

The Kosovo dead are the final tally to be added to the other atrocities of the 1990s: 250,000 lives lost in the former Yugoslavia, nearly a million lives lost between April and July 1994 in Rwanda, and dozens of other continuing conflicts throughout the world.[1] According to one estimate, those killed in the twentieth century in war and revolution number some 190 million people. That's equivalent to killing one out of ten people living throughout the world at the century's beginning.

Closer to home, it is not state-sponsored murder that we fear. Instead, in a country of unprecedented affluence, political stability, and the rule of law, we most fear being killed by a stranger, getting caught in the wrong place at the wrong time. That fear isn't unjustified. In fact, between 1972 and 1992, the number of murders committed each year by strangers doubled. Even more disturbing, the faces of those killers are becoming younger and younger. The most recent images to flood our national consciousness invoke places like Santee, California; Littleton, Colorado; and Jonesboro, Arkansas; and names like Andrew Williams, Mitchell Johnson, Andrew Golden, Luke Woodham, Michael Carneal, Eric Harris, and Dylan Klebold.

It isn't just the magnitude of these tragedies that propel them into prominence. Having become desensitized to years of images of youths in urban war zones, what is shocking about these latest crimes is that the perpetrators are mostly white boys in suburbs, small towns, and rural communities, the sorts of places thought of as refuges from urban crime, where families were living the so-called American dream. The larger truth is that the number of murders committed by youths has soared 168 percent since the mid-1980s:

- Over the past decade, rates of interpersonal violence have increased among young people, and the age of victims and perpetrators is getting younger and younger.
- Homicide rates among young males (thirteen to twenty-two) are increasing most rapidly. The tendency is increasing for their victims to be strangers or casual acquaintances.
- The lethality of interpersonal violence among young people has grown and appears to be associated with greater access to firearms, increases in gun caliber and in the number of automatic and semiautomatic weapons, or a greater willingness to use firearms.[2]

It is difficult to find agreement on explanations for the violence. Is there a gene for violence or a "homicide module" in the brain?[3] Is violence an instinctive drive that the coercive forces of civilization must keep in check? Or is the violent nature largely created by civilization, as psychiatric criminologist James Gilligan argues? Does it arise primarily from the workings of the male brain, or does the fact that one in four adolescent arrests for violent crime is now a female hold an important lesson?

In this chapter, we look inside the dark corners of the brain in search of answers to these troubling questions. It is a journey that is often uncomfortable, best read with all the lights on, but it is also one of promise. The emerging science of the biological roots of violence promises to go beyond the old dichotomies implicit in the questions we just posed, revealing the subtle and complex interplay between brain and world that triggers some of humanity's darkest deeds.

From Tanzania to Central Park

In the opening scene of Stanley Kubrick's film *2001, A Space Odyssey,* a protohuman ape in the midst of a group clash wraps his hand around a discarded bone and slowly raises it high over his head, to the dramatic accompaniment of Richard Strauss's *Also sprach Zarathustra.* A second later he crashes it down on another protohuman's skull, crushing it. In this rendition, that elemental act, the first act of war, is the dawn of human destiny. As that ape victoriously tosses the bone into the air, it transforms into a spacecraft, visually linking the violence of war to the unfolding of human history, where the music shifts to Johann Strauss's statelier "Blue Danube" waltz.

As intuitively appealing as this link appears from the perspective of history's bloodiest century, the idea that war is a distinguishing feature of humans was shattered by the events of January 7, 1974, in a remote corner of Tanzania. On that day, Hillali Matama, the senior field assistant at primatologist Jane Goodall's research center, was struggling to keep pace with a group of eight chimpanzees. Chimpanzees are normally a loud bunch, but these eight were traveling silently, as though purposefully, toward the edge of their range. Soon, they crossed out of their range and into that of the neighboring Kahama chimpanzee group. A moment later, the eight spotted Godi, a male member of the Kahama group, eating alone in a tree. When Godi saw them, he turned to flee, but not fast enough. The eight chased him until a large male, Humphrey, grabbed Godi's leg, sending him crashing to the ground. Humphrey jumped on Godi and pinned him down. The others attacked the now defenseless Godi, biting him and beating him over and over again. Hooting, they beat and bit Godi for a full ten minutes. As the beating continued, a female, Gigi, circled the group, screaming wildly with excitement. As Godi lay there, his face a pulpy mess, his body deeply bruised and covered with gaping punctures, one of the eight picked up a large rock and hurled it at him. And then Godi's attackers left him and turned deeper into Kahama territory, charging wildly before returning to their range. Godi was last seen limping away from the scene, surely to die shortly afterward from his wounds.

Hillali Matama had just become the first human to witness another species engage in a raiding party, an elemental act of war.

It is a brisk April night in 1989. A twenty-eight-year-old female investment banker from Salomon Brothers joins other hardy joggers around New York City's Central Park for some exercise. Unknown to her, a group of six youths, aged fourteen to sixteen, is also in the park. They are there to "have some fun," as they put it, and soon begin the search for suitable targets. As the jogger passes through a grove of sycamore trees, her path crosses that of the six youths. They corner her. Like Godi, she is in the wrong place at the wrong time. One of the youths, seventeen-year-old Kharey Wise, pins down her feet while a friend begins to slash her with a knife. Another beats her with a metal pipe while another picks up a rock and begins to smash her with it. For half an hour the attack continues, as they gang-rape her. All the while "everyone laughed and leaped around," as one of the attackers described it. Finally, when their victim no longer appears to be breathing, the attackers stop. Amazingly, although left brain-damaged and in a coma for weeks, the jogger did not die.

The Evolution of Aggression

At first glance, the similarities between these two incidents are striking. As Freud suggested, perhaps we share an aggressive, predatory drive with other species, one that can rise to a murderous frenzy. For Freud the horrors of World War I shaped this dark image of humanity; for another generation, it was the atrocities of World War II. The view's most famous voice was Nobel laureate Konrad Lorenz, the pioneering scientist of animal social behavior who is best remembered by the ducklings that followed him wherever he went. Lorenz argued that we share a basic aggressive instinct with other animals. Like Freud's theory of sexual energy as an uncontrollable force that builds until it spills over, Lorenz's theory held that your aggressive drive builds until it finds an inevitable release.

Today, the notion of an aggressive drive is out of fashion, yet that doesn't mean that biology plays no role in shaping aggression, that all aggression is learned. The selfish gene view of evolution, for example, suggested that the aggressive behaviors we see all around us—war, spousal abuse, homicide, rape—may be the result of natural selection. If what matters in the end is maximizing genetic fitness, then any strategy, including aggressive ones that maximize fitness, could be selected. Remember from chapter 1 the spotted hyena's den, where newborn twins' first worldly act is a death match with each other. Without conscience or remorse, evolution blindly makes survival machines. If your nature is a fallen one, it is because evolution doesn't make angels.

Some evolutionary psychologists are thus quick to point out that the !Kung San, an African tribe often depicted as naturally innocent and gentle, have a murder rate exceeding that of urban Detroit. Other evolutionary psychologists contend that rape is a genetically developed strategy sustained over generations of human life because in some situations it can be a successful reproductive strategy.[4] In particular, according to the "young male syndrome" view of violence, aggression is a fitness strategy for young males between the ages of fifteen and twenty-nine. Evolutionary psychologists further suggest that youth violence may be a risky competitive strategy that accompanies surges in muscle strength, aerobic capacity, and quick energetic bursts that might be needed for risky forms of aggression. Just look at the prison population, they argue. As evolutionary psychologist David Buss concludes, "The proportion of young males [aged fifteen to twenty-nine] in a population may be the best, or one of the best, predictors of violent aggression."[5]

There is one thing wrong with this conclusion.[6] Males aged fifteen to twenty-nine aren't the most aggressive age group. As Richard Tremblay of the University of Montreal remarks, "humans are more physically aggressive between the ages of twenty-four and thirty months than at any other time in their lives." According to Tremblay, the chief researcher of a long-term study of childhood aggression, the typical two-year-old engages in eight to nine aggressive acts an hour.[7] The consequences of a toddler's aggression typically aren't as dire as those of violence committed by someone of more years and means, but in terms of sheer numbers of aggressive acts, the toddler wins easily.

Pondering the implications of cultural biology for how we understand the nature of violence, we realized that young children's relative lack of self-regulation held an important clue. We didn't think it meant that children are simply wild, asocial beasts that need to be coerced into submitting to a culture's influence. We've noted in the previous chapter that children are born with prosocial behaviors and with an ability to empathize with others that primes them to socialize to the norms of their culture, but their prefrontal cortex, which underlies both their behavioral inhibition and their user's guide to life, is immature. Propelled by social desires, the capacity for self-regulation itself develops in accordance with cultural conventions as your prefrontal systems develop. Daniel Goleman called this your emotional intelligence and documented the crucial role that experience with competent adults plays in shaping a child's emotional intelligence.[8] The social competence that emotional intelligence confers is far more important for obtaining social status than aggressive strategies. This is true even in chimpanzee groups, where social status depends on forming coalitions, not pursuing aggressive strategies.[9]

Hence, young children act on their aggressive impulses in part because their behavior isn't yet under the guidance of their user's guide to life. During the course of normal development, the typical child learns how to inhibit this aggressive behavior and replace it with cooperative and peacemaking strategies, which lie at the core of social competence.[10] From the perspective of cultural biology, this suggests that one path to violence may lie in veering from this normal journey. In previous chapters, we explored how the key to unlocking the impressive capacities of your brain lies in extending and staging its development; thus the long period of maturation into adulthood. This came about by a process of progressive externalization as hominid brains exposed more of their developmental program to the world, letting the interaction between developmental mechanisms and a structured world shape the brain. Who you become depends not just on your genes and your environment in the womb, but equally on your experiences during childhood. We've seen why this is an immensely powerful strategy. But it also carries a potentially huge risk, a trade-off that we think is at the root of much violence. When the journey to a fully developed person derails, what changes resculpt the brain to make the normal balance between the internal guidance system and the user's guide to life go out of kilter?

A Tempest in the Brain: Your Faustian Bargain

You met Rachel in chapter 7. As her case and thousands of others reveal, patterns of neglect and abuse can literally resculpt the developing brain, often sabotaging the journey to personhood. Emotional interactions with a caring adult underlie the emergence of the child's sense of self, social competence, and capacity for empathy, helping to set up the neural circuits that shape the social dimensions of your personality and even how you respond to stress.

As we looked into the brain systems underlying violence, we were struck by the fact that noradrenaline and serotonin, two key chemicals in the internal guidance system, also play a central role in regulating violent behavior. The noradrenaline system has its roots in two small areas associated with the brain's flight-or-fight response. If you or someone you know has ever suffered a panic attack, this is the system involved. This response sets off a remarkable series of changes inside of you. It sparks your feelings of fear and anger; it increases your heart rate, elevates your blood pressure, even shuts down the blood supply to your intestines, diverting it to your muscles for quick action. We wondered whether toxic childhood experiences resculpt this system, and, if so, what this meant for understanding aggression.

There is abundant evidence that high levels of childhood stress alter levels of noradrenaline, a finding that has ominous overtones. We have found that your internal guidance system colors your world with positive and negative value and sends this information to higher brain areas to mediate decisions. This suggested that a child whose toxic experiences have resculpted his or her noradrenaline system might live in a relabeled world. The result would be a brain that sees danger signals everywhere and sends errant alarms to higher areas. Where your brain sees a smiling face, this resculpted brain might see a threatening smirk, setting the natural defense condition to a permanent state of emergency.

Could similar resculpting occur in the serotonin system? Serotonin is involved in our sociability and impulse control. A number of studies reveal that brain serotonin levels depend heavily on the environment.[11] In particular, Stephen Suomi and his colleagues have shown that lack of maternal care in nonhuman primates leads to reduced levels of serotonin, greater aggression, more alcohol consumption, and more anxietylike behaviors during adolescence. In fact, low levels of serotonin have been implicated in the development of alcoholism.[12] Its lowered levels in the stressed brain make for chemical trip wires for extreme and unpredictable violence.[13] Low serotonin levels also spark a cycle of aggressive behavior and social illiteracy that sets up the child for school failure and peer rejection.

Brains exposed to stress also appear to react differently to drugs. Neglected and abused children are especially at risk for future drug abuse, which begins perhaps as self-medication to alleviate the hostility, depression, or suicidal ideas that fill their minds. But consider what happens when monkeys exposed to high levels of stress are put back into a larger group. When researchers gave all the monkeys in the group a low dose of amphetamine, monkeys raised in standard environments were relatively unaffected by the drug. The neglected monkeys went berserk, attacking and killing others in the group.

These findings made us realize just what a dangerous gamble human brain development can be. On the one hand, our developmental dependence on the world holds many of the keys to being human; on the other, it contains lurking vulnerabilities that can derail the journey to becoming who we are. In addition to these dangers, there is another vulnerability that might be even more menacing.

Stress can ravage our bodies in many ways, but its effects on the brain can be just as deadly. Recent research has found that chronically high levels of stress trigger the release of highly toxic amounts of brain chemicals called corticosteroids, which at high levels literally attack brain cells, causing permanent damage. The truly menacing nature of corticosteroid damage becomes evident when we saw where in the brain it has its damaging effects. High levels of corticosteroids attack the hippocampus, which underlies long-term memory formation and retrieval, disrupting both memory and learning. These effects might also affect our ability to recall highly traumatic events, and so might lead to the amnesia that accompanies post-traumatic stress disorder. If stressful experiences continue over many days or months, memory and learning disruptions might take a more severe form by disrupting the ability to integrate one's experiences around a coherent and enduring sense of self. Since this ability lies at the core of creating a user's guide to life—and its rich repertoire of symbolic selves—these disruptions could have devastating consequences. Just think of what it would be like to be unable to associate your current behavior with your future self: Would the consequences of punishment worry you if you didn't realize that the *you* committing the crime would be the *you* facing the punishment?

Known clinically as dissociation, this inability produces a lack of connection in a person's thoughts, memories, feelings, actions, or sense of identity. In 1994, dissociative identity disorder replaced multiple personality disorder in the psychiatric diagnostic manual, reflecting changes in professional understanding of the disorder that resulted largely from increased empirical research of trauma-based dissociative disorders.[14]

As many as 99 percent of individuals who develop dissociative identity disorder have documented histories of severe and prolonged trauma during childhood. This astonishing figure prompted us to look deeper into the developmental roots of violence. In particular, since cultural biology emphasizes brain-environment interaction, we were interested in knowing whether a toxic environment is all that's needed to derail the journey to the person, or whether biological differences can put some people at risk. We wondered about this because we knew that we are all born with an "emotional thermostat," the temperament we examined in chapter 6. These characteristic ways of responding to the world reflect the neurochemistry of your internal guidance system, and they play a role in determining how people will interact with you. If you were a gregarious, happy child, chances are you elicited more positive interactions from others than did a sullen, reserved child. These temperamental styles could thus make the journey to the person either smooth or bumpy. Perhaps these styles also contribute to differences in the way children react to a challenging environment: One may be resilient, the other vulnerable.

This is a sensitive issue, because it can be misused to shift the blame of abuse. But we think it is important to look our biology squarely in the eyes because an extreme environmental view of other mental diseases has had a devastating effect on people's lives. For example, parents were once blamed for causing schizophrenia, a misdiagnosis that not only created a great amount of undeserved blame but also likely delayed the discovery of treatments. An extreme environmental view might likewise obscure important treatments for violent personalities. In some cases, biological vulnerabilities may overwhelm even the best parental practices. Consider, for example, how fifteen-year-old Kip Kinkel spent his last free hours in quiet Springfield, Oregon. After shooting his father in the head with the semiautomatic rifle his father had given him, he talked to his friends on the phone about the night's upcoming TV shows, wondering when his mom would get home. When she did, he shot her, too, then sat down to watch the shows. Sometime during the night, with his parents' bodies downstairs, he decided he'd shoot up his school in the morning, a decision that left twenty-two injured and two dead.

By all accounts, Kip's parents, Bill and Faith, both teachers, were nurturing and attentive parents who took their kids on hiking trips and family vacations and had close connections to their quiet neighborhood. But Kip seemed out of control. He bragged to his friends about his bombs—police found more than five stashed away in a crawl space in his home. In his literature class, he reportedly read from his diary about killing everybody and complained of voices in his head. He was fascinated by guns and liked to torture animals. His first brush with the law was for throwing rocks off a freeway bridge; his second was for stashing a stolen gun in his school locker.

What sort of disturbance could have been going on inside his brain? Beyond the knowledge that he suffered from severe depression and fits of anger for which Prozac had been prescribed, the rest remains a mystery. Kinkel's brain likely represents the extreme end of a biological predisposition toward violence. At the other end of the spectrum, children who emerge unscathed by even severely toxic environments, who are highly resilient, may be predisposed with brain chemistries that buffer them from the effects of such environments. Better understanding of these differences is an important goal that will lead to more effective intervention strategies.

The Brain of a Sociopath: Natural Born Killers?

As we saw in chapter 5, when the neurologist Antonio Damasio showed emotionally laden pictures to Elliot, his patient showed no bodily reaction as measured by a galvanic skin response, the sort of response used to uncover anxiety in lie detectors.[15] Although Elliot could describe the sorts of reactions he *should* have to such pictures, he was unable actually to feel them. This is really extraordinary when you consider the emotional reaction most of us have to images of accidents and other tragedies. In fact, it is common for jury members to become nauseated after seeing photos of crime scenes and to suffer nightmares long afterward.

Damasio showed that emotion and reason, far from standing in opposition to one another, are tightly intertwined in our decision making, which requires accessing the emotional significance of our experiences and memories. Without this capacity, Elliot and the dozen or so other patients who are under Damasio's care with a similar condition are incapable of planning for the future and are deficient in judgment and moral insight. It is as though their brain damage destroyed "what their brains have acquired through education and socialization."[16]

Elliot's tragic condition has provided a window into the brain centers where emotional significance is intertwined with experiences and memories. This area is the orbitofrontal cortex, a prefrontal region lying above the eye sockets. As we shall see, it has chilling implications for understanding violence.

Damasio called the condition of this group of patients "acquired sociopathy." As the name suggests, many of its symptoms are analogous to developmental sociopathy. Indeed, Damasio and his colleagues have recently explored whether children who suffered brain damage to the same regions as did Elliot have disturbances in social behavior.[17] They examined this question in a twenty-year-old female who suffered brain damage as the result of an accident at the age of fifteen months, and a twenty-three-year-old male who suffered brain damage at the age of three months. Both individuals, whose parents were highly involved in their upbringing, appeared typical for many years after their injuries. As they got older, however, their behavior progressively worsened, ranging from a lack of responsiveness to punishment, increasingly disruptive behavior at school, chronic lying, and sudden outbursts of anger, to petty thievery and reckless behavior. The woman's behavior required her to be placed in a treatment facility at fourteen, the first of several admissions.

As with Elliot, tests revealed that these two patients had normal intelligence but were insensitive to the future consequences of their decisions, had defective autonomic responses in the face of punishment, and didn't respond to interventions. There was one major difference: Elliot had a stockpile of moral knowledge that he had acquired before his brain damage, which he could call on but had difficulty applying to his situation. In contrast, these two patients had faulty moral reasoning abilities, suggesting that their early brain damage impaired their ability to acquire social conventions.

The disturbing behavior of Damasio's two patients is symptomatic of sociopathy, or what is more precisely defined as antisocial personality disorder. Sociopaths are manipulative, callous, impulsively aggressive, and show no remorse for the harm they cause others. They feel no empathy for others, or guilt, and form only the most superficial of attachments. In the extreme, this makes them the embodiment of pure self-interest, viewing people as disposable props for their own ends. Some estimates put the number of sociopaths between 1 and 4 percent: Most of these individuals are living in the general population. Their irritable, argumentative, and intimidating personalities cause others to avoid them, which leads to difficulty holding jobs, involvement in domestic violence, traffic offenses, and severe marital difficulties.

It isn't surprising that sociopaths run into problems with the law. Indeed, about 20 percent of the prison population is estimated to be sociopathic. Researchers have performed brain-imaging experiments with sociopathic criminals to discover how their brains respond to emotional images and words.[18] Strikingly like Damasio's patients, sociopathic criminals have no emotional reaction, and the part of the brain that is damaged in Damasio's patients appears less activated in these criminals compared with normal subjects who show large activations in their orbitofrontal cortex in response to the same images and words. This is a striking convergence of findings, particularly as recent imaging of the human brain reveals that the orbitofrontal cortex is involved in your use of theory of mind; that is, when you think about other people as being persons like you.[19]

Imagine what it must be like to see someone in distress and feel nothing. Conversely, imagine someone looking at you and not seeing a person with feelings and hopes, but simply another object. Indeed, sociopaths may also be insensitive to the threat of punishment. For example, anxiety-related chemicals rise in most prison inmates as their trial dates near, but for sociopaths these levels stay flat. And whereas you would move your hands to avoid an electric shock, many sociopaths do not, because the specter of receiving a shock produces no anxiety in them.

Adrian Raine at the University of Southern California and his colleagues recently examined the brains of people with antisocial personality disorder who were living in the general population.[20] Raine and his team advertised for volunteers for their study at temporary employment agencies, since temporary workers had a relatively high incidence of violence, including homicides that had never been found out. After psychological screening, twenty-one men were diagnosed as suffering from antisocial personality disorder, twenty-seven men were diagnosed as suffering from substance dependence, and thirty-four men were found to have neither antisocial personality disorder nor substance dependence (the control group). Brain scans revealed that the antisocial personality disorder group had an 11 percent reduction in prefrontal gray matter volume compared with the control group and a 13 percent reduction compared with the substance abuse group. When asked to talk about their faults, a normally socially stressful experience, the antisocial personality disorder group showed significantly less autonomic response than did the two other groups.

This was the first study to reveal brain differences in people with antisocial personality disorder in the general population. As with the cases Damasio examined, it raises the question of whether such individuals are made or born.

As we considered this troubling question from the perspective of cultural biology, we saw that the question rests on a false dichotomy: There isn't an either/or answer. Instead, the answer lies along a continuum of brain and external influences that converge to create the individual. Some have speculated that sociopathy is simply one of the personality types that evolution has selected for, a personality that succeeds by manipulation and lack of remorse, whether in a criminal or a driven executive.[21] From our perspective, this explanation is too simple, since personality is a complex blend of temperament and life experiences. In this regard, consider the view of David Lykken, a strong advocate for the genetic view of human behavior. Despite his belief in the dominance of genetics, Lykken thinks that most sociopathic behavior is the result of faulty parenting, a failure to socialize children.[22] Perhaps Kip Kinkel and others like him represent an extreme case, in which early behavior makes them exceptionally vulnerable to sociopathy despite intensive parenting efforts. Perhaps many others have a temperament that can be influenced by socialization to become successfully integrated into society.

As we considered the role of socialization in the genesis of violent personalities, we were struck by the findings of James Gilligan, director of the Center for the Study of Violence at Harvard Medical School and former director of mental health for the Massachusetts prison system. Having spent years inside prisons studying the mental health of inmates, he remarks that as children these men "were shot, axed, scalded, beaten, strangled, tortured, drugged, starved, suffocated, set on fire, thrown out of windows, raped, or prostituted by mothers who were their 'pimps.' "[23] As a means of coping with their own anguish, many of these children never learn to empathize with the pain and suffering of others. As Gilligan suggests, their deadening of feeling combined with the shame of their social rejection snares them in a vicious cycle of violence. Even for those who can still empathize, their immersion in neglectful, violent worlds makes them prone to impulsive behavior, letting trivial events spark the tragedies that are the all-too-common fare of daily news.

Gilligan also notes something that is often overlooked in describing inmates in maximum-security prisons. Although we may think that most injuries in prisons come from fights with other inmates, Gilligan notes that most injuries are self-inflicted. Gilligan describes how many inmates have not only slashed their bodies or swallowed razor blades and other sharp objects but also cut off their penises and testicles, and even clawed out their eyes.

Why would so many men in prison mutilate their own bodies? Recall that the orbitofrontal cortex integrates emotions with our memories and experiences. The loss of this capacity results in emotional numbness, and indeed many inmates use metaphors of emotional death to describe themselves. For these men, then, feeling pain from severe trauma is better than feeling nothing at all.

As we explored the roots of violence, it became evident that it would be a mistake to suppose that all forms of violence can be reduced to the journey of a person interrupted. A unifying theory of violence with a single root cause would be an oversimplification. Violence has multiple roots, and so our next step was to explore another form of violence that doesn't spring from interrupted socialization, but paradoxically from the forces of socialization themselves.

The Killing Crowd

In 1989, as the Soviet Union fell apart and the cold war that dominated a generation came to an end, many predicted a new era of political stability. Indeed, the social scientist Francis Fukuyama captured a great deal of attention when he suggested that this was the end of history—not, of course, the end of time, but the culmination of history's progress in liberal democracy, which would spread throughout the world as the final form of government.[24] Recent history has, however, told a different story. Indeed, the end of the cold war seemed to rekindle latent ethnic ties and identity, and, if anything, has undone modern ways in many parts of the world. In its wake came renewed ethnic strife and cultural clashes, suggesting that globalization had been preempted by Balkanization.

With the return to ethnic identities came cultural wars and the horrors of genocide. Do these images from Bosnia, Kosovo, and other locales prove the fragility of civilization and reveal our individual instinct for aggression writ large?

There is a problem with the idea that warfare or systematic killing by groups is rooted in our instinctive aggression. Systematic killing by

groups is in a real sense a social activity, one requiring an extraordinary amount of cooperation among people. Wouldn't the real consequence of the breakdown of social order be anarchy, the war of all
against all that Hobbes envisioned? In the examples of violence we
have examined so far, from the impulsive to the serial killer, the perpetrators commit their acts on their own. In fact, the serial killer is the
icon of the loner, who lives on the margins of a social order. Moreover,
systematic killing by groups often comes about by the chillingly rapid
transformation of peaceful individuals into killers. What changes
might take place in the brain to make this transformation happen so
quickly?

Consider the story of Reserve Police Battalion 101. The five-
hundred-member troop were middle-aged family men, between thirty-
three and forty-eight years old, of working- and lower-middle-class
backgrounds from the city of Hamburg (the middle age of these men
conflicts with the "young male syndrome" view of violence). None of
Reserve Police Battalion 101 had prior military service, only five were
members of the Nazi party, and none was an SS member. These
weren't young men who had grown up on Nazi ideology.

In the summer of 1942, Reserve Police Battalion 101 was sent to
the Polish village of Józefów. Their instructions were to round up 1,800
Jews, pick out the men of working age for shipment to a camp, and
then kill the women, children, and the elderly. Before they started their
work, however, their commander told them that any men who didn't
feel up to the task could step out. Only twelve members did so.

As the gruesome task continued, soon others began to quit, particularly those who couldn't master the "neck shot" that was supposed to result in a clean wound, but instead splattered them with their victims'
blood and brains. Yet in all, only 10 to 20 percent of those assigned to the
firing squads sought release or evaded the shooting by more surreptitious
methods. Indeed, as witness reports later testified, often there were so
many volunteers that some had to be turned away. Another policeman
remembered that musicians and performers in a visiting entertainment
unit of Berlin police "asked, indeed even emphatically begged, to be al-

lowed to participate in the execution of the Jews." In less than a year's time, this battalion had murdered at least thirty-eight thousand Jews and sent another forty-five thousand to their deaths in Treblinka.

The historian Christopher Browning has painstakingly reconstructed the story of Reserve Police Battalion 101 from the investigation and prosecution of 210 of the battalion's members conducted between 1962 and 1972 by the office of the state prosecutor in Hamburg.[25] The most obvious potential explanation for their repugnant behavior is that these men were coerced, either by implicit or explicit threat of punishment. But, as Browning notes, "in the past forty-five years no defense attorney or defendant in any of the hundreds of postwar trials has been able to document a single case in which refusal to obey an order to kill unarmed civilians resulted in the allegedly inevitable dire punishment."[26]

Drawing on group behavior research, Browning concluded that the members of the battalion were ordinary men who, through the combined forces of peer pressure, conformity, and the dehumanization of the enemy, were induced to commit what might seem to us to be unthinkable acts. Of course, by trying to explain the atrocities of Reserve Police Battalion 101, Browning is by no means excusing them. But if this group of mainly middle-aged family men with no military experience could so quickly corral, cull, and kill, we wonder what group of men cannot. Indeed, Browning ends his study by suggesting that far from a flight from civilizing forces, it is these very forces of civilization that make killers out of ordinary men.

We have seen this sort of explanation before in Philip Zimbardo's prison experiments. We wanted to know whether looking inside the brain could help us understand this dangerous transformation of ordinary men into group killers. UCLA's Itzhak Fried has suggested that a common set of symptoms characterizes the rapid transformation of previously nonviolent individuals into group killers. These symptoms constitute what he calls "Syndrome E," which has at its roots a cognitive fracture characterized by obsessive ideation, compulsive repetition, rapid desensitization to violence, blunting of emotional response, and hyperarousal, together acting like a contagion that spreads through the whole group.

Since this syndrome has so many social qualities, it seems implausible that it stems from a primitive aggressive force. Instead, its social qualities suggest that a hyperarousal of the orbitofrontal cortex may underlie it. Whereas the sociopathic behavior of isolated killers involves diminished activity in the orbitofrontal cortex, systematic killing by groups seems to tell a strikingly different story. Far from the emotionally deadened loner who can't maintain social bonds, a sentient person immersed in a social world of obsessive ideas may experience a tempest inside the orbitofrontal cortex that rearranges the emotional significance of his or her experiences. Since the orbitofrontal cortex sends massive projections to subcortical centers, when it is hyperaroused it may interfere with emotional regulation, disconnecting action from appropriate emotional meaning. Consider the disconnection in the angry Palestinian mob in October 2000 that lynched two Israeli soldiers, holding their bloodstained hands high in the air as they screamed victoriously, or the men dancing around the battered body of truck driver Reginald Denny as the Rodney King verdict tore through Los Angeles on April 29, 1992. More recently, crowds of Palestinians danced in the streets at the news of the September 11, 2001, attacks.

Perhaps the orbitofrontal cortex becomes hyperaroused by a group contagion of ideas that overrides your normal capacity to regard others as human like you. If so, this might underlie the connection between group killing and ideology. We doubt group killing could occur without obsessive ideation, since an essential step in targeting a group for genocide is a dehumanizing ideology. The real possibility that cultural ideas can stunt empathy and reset the emotional values we attach to other people is an ominous one. This is the dark side of cultural biology: Far from explaining violence as a primitive reflection of our nature, group killing would be an extraordinary case of culture feeding into our biology to reshape behavior.

In this scenario, group killing is the most recently evolved, most human part of us out of control. Group violence doesn't stem from the breakdown of society, but flows from the same brain circuits that normally drive socialization. Although the modern image sees our more recently acquired brain structures as sitting on top of ancient centers of aggression, we realized that this image needed to be turned on its head. Much of what is disturbing about human behavior doesn't spring from the broken restraints of ancient parts, but from the vast veto power of the new parts. Your great sociability and capacity to internalize beliefs can result in a dangerous hyperarousal of the very parts of the brain that distinguish you as human, paradoxically leading you to dehumanize others.

From the Robber's Cave to Littleton

The process of how dehumanization occurs is enormously complicated, but some striking clues emerged in the 1950s in an unlikely place: a boys' summer camp near Robber's Cave, Oklahoma. Twenty-two middle-class eleven-year-old white Protestant boys, all well-adjusted—as alike as possible—were invited to spend three weeks at summer camp. The boys had been selected from different schools, so no one knew anyone beforehand. Unknown to them, their "counselors" were part of a research group headed by the University of Oklahoma's Muzafer Sherif, who had organized the summer camp as a social experiment. The psychologists split the boys equally into two groups before meeting, and each group was sent separately to a different part of Robber's Cave State Park, so that they wouldn't know about the other group's presence at first. One group decided to call themselves the Rattlers, while the other dubbed themselves the Eagles. The plan was that after about a week the two groups would enter into athletic competitions—baseball, tug-of-war, and other typical camp fare—with the prediction that hostility would emerge.

Sherif and his colleagues didn't have to wait until the athletic games for hostility between the Rattlers and Eagles to break out. One day the Rattlers heard the Eagles playing nearby and wanted to run them off.

Now aware of each other's presence, the groups implored their counselors to let them compete against each other. The first game was baseball, which started with name-calling between the two groups. The Rattlers decided to hang their flag over the back of the baseball diamond, since they felt that the diamond was really theirs. The Eagles, who lost the game, were so incensed that they tore down the Rattlers' flag and burned it.

Soon, the groups were raiding each other's cabins, ripping apart their contents, and stealing things. One day the Eagles decided on a daytime raid and carried sticks and baseball bats in case the Rattlers were at home. Piles of rocks and socks filled with rocks were stockpiled in defense in each cabin.

Sherif and his colleagues decided they had better move on to the next stage of their experiment: reconciliation. They began with strategies such as common mealtimes, but this only led to enormous food fights. Finally, they moved the boys to a new location, but this hardly eased tensions. It was only when they created a scenario in which outside vandals were destroying the camp's water supply that the two groups allowed an uneasy truce.

Many factors complicate a simple interpretation of what happened at Robber's Cave.[27] For example, the fact that the two groups were allowed to form strong within-group bonds first may have contributed to their between-group hostilities. Nonetheless, an enormous amount of evidence confirms our readiness to form groups, to categorize people on the basis of their group membership, and to favor our own. Known as ingroup-outgroup bias, this is a strikingly pervasive human characteristic. It's no accident that home team and visiting fans at college games are typically seated on opposite sides of the field.

When we reflect on the crucial role that collective identity—the public self—plays in coordinating a group to solve problems for mutual benefit, it's not very surprising that it could have these dark consequences. What remains to be discovered, though, is whether the ingroup-outgroup bias is inborn and thereby prompts us to form groups, or if this bias is a side effect of a more basic readiness to form group identities.

What happens inside the brain when you shift your working self-concept from private self to public? Do you recode your emotional experiences to be in line with group sentiments? Perhaps your private emotions and moral judgment become invisible to you. Maybe the hallmark of sanity is the ability to maintain both your private self and a public one, never becoming so lost in a group that you lose your individual values and opinions, or so isolated that you feel no bond to others. Consider the following. A year before the Columbine High School massacre, Rachel Scott was performing in a talent show when her music cut out. Up in the sound booth, a student rushed to hook up a reserve tape deck to save her performance. The student was Dylan Klebold, who a year later walked up to her with his friend Eric Harris and shot the seventeen-year-old in the head in cold blood. What profound changes took place inside Klebold's brain to result in such a disturbing change in his behavior?

Were Harris and Klebold involved in a dehumanization campaign that recoded their brains' value systems? The most obvious answer stems from comparing the typical high school to Robber's Cave. In any typical high school you can find divisions analogous to the Rattlers and the Eagles. They're called jocks, preppies, druggies, Goths, skaters, and nerds. In high school, life is a time of such uncertain personal identity that group identities are all the more salient, as are media images that hold up identities for youths to aspire to. Whether or not the Trenchcoat Mafia that Harris and Klebold were a part of was a tightly cohesive group, it is clear that they were part of the outsiders, and that they were incessantly picked on by the jocks, having bottles and rocks thrown at them from passing cars. Even worse, they had no option but to return every day to the same outgroup treatment. One Trenchcoat had brandished a shotgun at a group of jocks. The graffiti in the boys' bathroom included "All Jocks Must Die." It seems clear, then, that Harris and Klebold created a strong ingroup-outgroup bias from their social rejection, soon glorifying in their outsider role—a typical reaction motivated by the need for positive self-regard—and dehumanizing the jocks, one of the Syndrome E symptoms. They became enamored of Nazi culture, learned enough German to greet one another with

Nazi slogans, and chose the anniversary of Hitler's birthday to commit their crimes.

Consider another item on the list of Syndrome E symptoms: compulsive repetition. Harris and Klebold talked obsessively about whom they would like to kill. They made a videotape for a class in which they pretended to shoot all the jocks, using a real rifle. An English teacher was so disturbed by the level of violence in one boy's writings that he brought them to the attention of a guidance counselor. Harris kept a hate-filled website warning of ominous things to come.

As much as they might help uncover the seeds of the boys' actions, there is obviously a large gap between these behaviors and actually committing atrocities. The search for other factors must include looking deeper into the boys' state of mind. The dynamic between the two, for example, appears to have been a dangerous one. Eric Harris appeared to be more of a leader, but also more troubled, while Dylan Klebold appeared to be more of a follower, susceptible to outside influence. Consider also that the marines had a week earlier rejected Harris, the son of a decorated air force pilot, because he was taking antidepressant medication. In an analogous case, the Green Berets had rejected Timothy McVeigh, another angry misfit, shortly before he blew up Oklahoma City's Alfred P. Murrah Federal Building. Did this pattern of rejection further solidify an ingroup-outgroup bias in both Harris and McVeigh, motivating them to construct more grandiose schemes? Both Harris and McVeigh selected symbolic dates for their crimes—McVeigh, the anniversary of the government's raid on a cult compound in Waco, Texas; Harris, Hitler's birthday—further evidence of obsessive ideation.

There's no doubt about the immense power of social rejection to create a dehumanizing campaign. Yet there are lots of disgruntled teenagers who live on the margins of the social order. Why do some cross the line and make the deadly transition from vengeful fantasies to murderous action? One of the most chilling insights stems from a study by Lieutenant Colonel Dave Grossman, a professor of military science. Grossman studies a problem that military commanders have long faced, one that might appear surprising to anyone accustomed to the common image of men as being hyperviolent. Historically, the overwhelming majority of soldiers have been unwilling to fire their weapons in battle, even when their lives are in immediate danger. Indeed, during the trench warfare of World War I, informal truces were often made by the enlisted men, who climbed out of their trenches to play soccer with one another. During Christmas, both sides sang Christmas carols to the other from their trenches.[28] During World War II, for example, only 15 to 20 percent of soldiers actually fired their weapons while in combat. In Korea, the number rose to about 50 percent. But by Vietnam, the number was up to 95 percent. Grossman wanted to know what had changed in Vietnam.

Grossman argues that between Korea and Vietnam the army changed the way it prepared soldiers for battle by incorporating psychological principles into their training. One of the simplest but most effective differences Grossman cites is a change in target practice. Traditionally, soldiers practiced marksmanship by shooting at a stationary bull's-eye while lying in a grassy field. The Vietnam-era trainees, however, were put into foxholes in full combat gear. At random intervals, olive-drab, man-shaped targets would pop up briefly. The soldier had to aim quickly and fire, with hits recorded by the target's falling over. If the soldier didn't fire in that time, the target disappeared. As Grossman notes, "What is being taught in this environment is the ability to shoot reflexively and instantly and a precise mimicry of the act of killing on the modern battlefield."[29] In training police and military snipers today, it is common to use lifelike models as targets, even with their heads stuffed with a combination of cabbage and ketchup to simulate a head wound.[30]

Grossman points out that these modern military training techniques have an uncanny similarity to many of today's point-and-shoot video games:

> [T]he same tools that more than quadrupled the firing rate in Vietnam are now in widespread use among our civilian population. Military personnel are just beginning to understand and accept what they have been doing to themselves and their men. If we have reservations about

the military's use of these mechanisms to ensure the survival and success of our soldiers in combat, then how much more should we be concerned about the indiscriminate application of the same processes on our nation's children?[31]

It is no coincidence that the marines have customized a version of Doom, an immersive shoot-'em-up computer game, for their own combat simulations. This does not mean that the path from video games to violence is simple and straightforward. Searching for *the* cause of violence is as misguided as searching for *the* cause of a tornado. Both are the product of a web of forces that accumulate in subtle and complex ways until they reach some critical point. The forces that cause a tempest in the brain that leads to violence are complex, sometimes interacting with one another, sometimes multiplying their joint effects, sometimes canceling each other out. Indeed, it is likely that children like Eric Harris are drawn to these violent games. Once Harris was exposed, obsessive game playing interacted with the other risk factors present in his life to nudge him closer to acting on his fantasies. It is chilling to consider that Harris customized the rooms in Doom to simulate the physical layout of Columbine and that he played it over and over again in that mode.

There is something else about computer game play that Grossman does not discuss. Brain researchers have used the computer game Tetris to see what brain changes take place as you learn and master a new task. In Tetris the object is to rearrange falling objects—squares, straight bars, L-shaped bars—so that they form a wall without any gaps. As you first learn Tetris, it takes total concentration. Since the demands of the game are new, many different brain areas are called into action. After you have become skilled in the game through extensive practice, however, less and less of your brain is activated, particularly in the cerebral cortex. What has happened? The most likely answer is that your brain has devoted special circuits to the game, dedicated Tetris networks that probably have developed efficient strategies for the game.

This process is called "automaticity," the overlearning of a task to the point where it becomes reflexive. How many times have you pulled into the garage after some routine errand and realized that you have no recollection of having actually driven there and back? As Grossman points out, the goal of military training is automaticity, the overlearning of a skill to the point where it can be engaged reflexively. Seeing that the brain actually makes a transition from using high-level self-monitoring regions to dedicated, automatic ones for the task suggests how this is accomplished. And it suggests that violent video games are prime tools for making specialized killing centers. Again, this doesn't suggest that such games turn every player into a homicidal time bomb, but it does suggest that by transferring cognitive control of action to a reflex they provide a powerful enabling force for those already prone to violent action.

Cultural biology emphasizes that your behavior is under the influence of multiple systems. Often these systems are highly integrated and share their information to make a sound, consensual decision. Sometimes, however, tempests arise in the brain and disrupt this balance, transferring control of your behavior to one system. Just as we cannot precisely forecast the storms of other complex systems such as the weather, we have a long way to go before brain tempests can be predicted by analyzing the many forces that converge in complicated ways to create them. The most important lesson so far may be that no single factor is likely to be the cause of violence. Violence itself is heterogeneous, ranging from impulsive rage to the icy premeditation of a sociopathic killer to the social contagion of group killing. Perhaps nowhere is the intricate blend of cultural and biological forces more complete—and more ominous—than in the roots of violence.

Is there a solution to the problems of adolescent violence? Our biology certainly doesn't suggest the perfectibility of human beings. But in a debate flooded by statistics, there is one more worth considering. For every young person who turns his violent impulses on another, there is one who turns it on himself. Both teenage murder and suicide rates have escalated to alarming rates. Indeed, as Patricia Hersch documented so poignantly in her book *A Tribe Apart,* the life of adolescents in America is far more lonely than violent. The violence that some adolescents turn to is in part a symptom of the larger problem of how modern society structures adolescent life so that it lacks adult involvement at a time when their struggle to form a coherent self-image requires guidance. Some of them, overwhelmed by the challenge of navigating that adolescent world by themselves, create worlds of violent fantasy that spill over to shatter everyone's reality.

Inside 9
Intelligence

Rethinking What

Makes Us Smart

onsider the last hours of Rob Hall, one of Mount Everest's most experienced guides. Hall and his group were falling behind schedule when they reached Everest's summit on the afternoon of May 10, 1996. Looking down into the valley, Hall saw a storm brewing and hastened their descent. On the way down, one of Hall's clients, Doug Hansen, an American postal worker, was struggling and had to stop. He and Hall lay down to wait out the storm that now was threatening to blow them off the mountain. Sometime during the night, Hansen died of exposure. Members back at base camp feared the worst for Hall, but at 4:30 in the morning his radio crackled to life. Trapped alone on the south summit of Everest at 28,700 feet in hurricane-force winds that drove the temperature down to

−140 degrees Fahrenheit, Hall had one final request: for his radio to be patched through via satellite to his pregnant wife back home in New Zealand. They talked for several hours, trying to reassure one another, and even discussed names for their baby. All the while, a rescue team was struggling a mere seven hundred feet below to reach him. With a final "don't worry too much about me," Hall switched off his radio. The rescue team couldn't close the final distance and gave up hope. Shortly after, Hall succumbed to the savage conditions. In total, nine would die, making it the single worst tragedy on Everest.

As tragic as Hall's death was, his last hours also capture the profound change communication and information technologies have made to our world. In the worst possible conditions in the most remote place on earth, Hall could instantaneously communicate with his wife halfway around the world even though a rescue party couldn't traverse a mere seven hundred feet. In fact, millions were following the tragedy on NBC Interactive, an Internet site. Their correspondent, New York socialite Sandy Hill Pittman, had made it to the summit the same day and kept the world informed moment by moment through the $50,000 worth of communication tools she took with her.

In our own lives, we feel the repercussions of change all around us: in the end of the industrial age that relied on a mainstream of low-skilled jobs; in the turbulence of corporate reorganization; in an economy once defined by borders now melding into a global one. These massive changes, perhaps a major turning point in human history, pervade every facet of our lives as we enter a new century that is as much a symbolic transformation into a new world as the passing of one more year.

Some call this new world a knowledge society, as knowledge is the new competitive advantage and the knowledge worker its prized occupation. The pace of change in this emerging society attests to our brain's immense capacity to both shape the world and adapt to its changes. It's a basic truth that human evolutionary success came not from strength or speed but from mental prowess. Ironically, that prowess is being challenged by the increasingly complex demands of the knowledge society. Will we be smart enough, individually and collectively, to succeed in this society?[1]

In this chapter, we explore some of cultural biology's far-reaching lessons for rethinking what makes us smart. Cultural biology provides the foundation for a view of what we call "constructive intelligence." We believe that constructive intelligence reveals the inadequacies of the modern view of intelligence, which stems from early-twentieth-century observations. In this chapter, we show how the lessons of constructive intelligence apply to the challenges of today's knowledge society. Our first stop will take us back to the roots of the modern view of intelligence, showing why this view not only doesn't stand up to scrutiny in light of today's brain science but also is an obstacle to creating the cultures of intelligence required to meet today's unique challenges successfully.

Intelligence and the Modern Image

The rise of the modern view of intelligence began in 1904 with the French minister of education, who had a problem. French schools were expanding rapidly, and their teachers needed a practical way to identify children in need of special help. As the director of the Sorbonne's psychology laboratory in Paris, Alfred Binet was a natural choice for this task, and he proved to be a good one.

Binet responded with a collection of puzzles, which he hoped would act as a window on the mental abilities of children. Because he believed intelligence had many facets, Binet chose a wide scattering of tasks, everyday problems like identifying common coins, naming the months of the year in order, finding rhymes for words, and describing a picture. He was also careful to train assistants to administer the tasks interactively, one on one. For Binet, though, the tests were merely a first step toward intervention. They helped identify who needed extra help, and how much. In fact, Binet devised a successful program in what he called "mental orthopedics."

Buoyed by the program's encouraging results, Binet wrote optimistically about their potential as a practical solution to improving intelligence. While the attention his program was attracting excited him, however, Binet was also concerned. It was one thing for him to use the test under his supervision, but he worried about what would happen when its use was outside his control. He realized that he would have to establish standards for others to follow; otherwise, the interpretation of test results could vary wildly among different testers.

Binet devised an ingenious solution. At first glance, there doesn't seem to be an objective measure of the difficulty of a task like identifying the common coins of American currency. But one way to establish relative difficulty with regard to learning lies in seeing at what age a person can master a given task. If you applied this strategy to numerous tasks that most children master, you would be inventing a profile of the "typical" child. Binet soon had an entire arsenal of age-ranked tasks. When testing a child, he would start with the easiest task, the one the youngest could perform, and work up to the harder tasks. The point in the progression where the child failed would then reveal what Binet termed the child's "mental age." So, for example, a child of eleven who could only master a task performable by the "typical" eight-year-old would be categorized with a mental age of eight, putting him three years behind the "typical" eleven-year-old.

It was a monumental step. For the first time, educators had an easy-to-administer *standardized* test to measure children's mental ages. Binet's test came just as interest in individual differences was rapidly rising, opening up many new possibilities to measure and compare children. It wouldn't take long, after a little technical tinkering by the German William Stern, for the notion of an intelligence quotient, or IQ, to emerge.[2]

As the use of his test spread and he lost control over its administration, Binet began to see a disturbing trend that caused him much anguish. He realized that a test that was supposed to help eliminate differences through compensatory education could just as easily magnify and cement differences to drive deep divisions among people.

Binet's fears would soon be realized, not in Europe but in America.[3]

America Retools IQ

The first few decades of the twentieth century would witness more change, and more wholesale abandoning of old ideas, than perhaps in all of history before. As these jarring changes helped create our modern image, the notion of human intelligence also underwent a profound transformation. Revolutions struck virtually every facet of human experience, including science, work, politics, and education. Even the art world would reflect the "shock of the new," as one critic dubbed it. Revolutions work by first burning the palaces, either real or intellectual. And the revolutions that built the twentieth century were no exception. In science, a radical new quantum mechanics swept away the old deterministic worldview. While quantum mechanics was creating the science of the elementary constituents of matter, Einstein was revealing the deepest nature of time, space, and the universe itself.

Sweeping changes in technology and physics were matched by a revolution in biology. In 1900, three botanists, one in Germany, one in Austria, and one in Holland, simultaneously but independently redis-covered Gregor Mendel's work on inheritance, which had been lan-guishing in obscurity for more than thirty years. Just two years later, the English biologist William Bateson, who later coined the term *ge-netics,* published *Mendel's Principles of Heredity: A Defence,* which contained an English translation of Mendel's original study. With it, modern genetics was born.

Mendel's work provided the crucial missing piece in Darwin's the-ory of natural selection: an account of the physical mechanism by which parents pass traits to their offspring. In working out the patterns of inheritance in peas, Mendel argued that parents pass on a collection of discrete particles, each of which determined a trait. Mendel showed that during conception both parents contribute a particle for each trait, such as pea color. Some kinds of particles dominate the others, which then recede into the background—not gone, just dormant. For a re-cessive particle to find expression in the next generation, it would have to be matched with the same recessive particle from another plant. In his work on dominant and recessive genes, Mendel had discovered the fundamental laws and the mechanism of inheritance.

The rise of modern genetics and its union with Darwinism had a profound effect on both science and society. Against the backdrop of modernism, Darwinism was reinterpreted as scientific proof of progress. Life didn't sit still; it progressed by evolution. Darwinism itself now pointed to the Utopian world that science and technology were making in the America of the early twentieth century. With Darwinism vindicating the notion of progress, another possibility gained appeal: Human intervention could speed along progress. Francis Galton, Darwin's own cousin, had earlier championed this idea and dubbed it "eugenics," the advancement of humanity through controlled breeding. In his 1869 work *Hereditary Genius,* Galton included a chapter, "The Comparative Worth of Different Races," in which he argued that intelligence was both largely inherited and different among groups. The evolutionary advantage of certain groups could be fostered, Galton maintained, both by "positive eugenics," the selective breeding of the favored class, and by "negative eugenics," curtailing the breeding of lowly classes.

In the United States, the union of Darwinism and genetics propelled eugenics into the mainstream of scientific thought. For example, every founding member of *Genetics,* the field-leading scientific journal, was a supporter of eugenics.[4] In 1906, the American Genetic Association formed a Committee on Eugenics, headed by Stanford University president David Starr Jordan and Charles Davenport, the eminent Harvard zoologist. Davenport would go on to form the Galton Society, an inner circle of elite eugenicists, including luminaries such as Edward Thorndike, the founder of American educational psychology.

At the core of eugenics in the United States was the straightforward connection between single genes and major human traits. This led to some odd claims. Davenport, for example, believed that thalassophilia, love of the sea, was a sex-linked recessive trait because he encountered it only in male naval officers. Davenport also thought that there were grave physical dangers in the interbreeding between tall Americans and short immigrants from southern Europe. If a child received the gene for height from the tall American but the gene for internal organ size from the short immigrant, he postulated, the mismatch would mean the children would have internal organs that were too small for their body size. It was a bit like the exchange between the dancer Isadora Duncan and George Bernard Shaw, himself an early supporter of eugenics. She once remarked to him that they should have a baby together, combining his mind with her body. Shaw replied that he feared the poor child might end up with his body and her mind.

Larger currents of change also were at the roots of the eugenics movement in the United States. In particular, an unprecedented flood of immigration was changing the face of America. Under the watchful eye of the Statue of Liberty, dedicated in New York Harbor in 1886, over fifteen million newcomers arrived by 1920, more than the entire U.S. population of 1850. Unlike earlier waves of immigration from northern Europe, the new immigrants were from southern and eastern Europe: Italians, Jews, Poles, Hungarians, Russians, and others.

Well before the rediscovery of Mendel's work, eugenicists were calling for immigration restrictions. From a legal stance, American immigration laws, which first came into existence in 1875, placed no numerical quotas on immigration prior to 1921. So immigrants couldn't be excluded simply on the basis of their country of origin. Certain "undesirable" categories were excluded, however. By 1907, "feeble-minded persons" were added to the list of undesirables. Feeblemindedness, though, didn't mean just a lack of smarts; it also carried heavy moral overtones including criminality and sexual immorality.

Eugenicists had no doubt that feeblemindedness was a Mendelian trait. For a while, feeblemindedness was screened by visual inspection—a trained look into the eyes of a feebleminded person would supposedly reveal the telltale vacant stare. This, though, lacked scientific rigor. Henry Goddard, director of research at the Vineland School for Backward and Feeble-Minded Girls and Boys in New Jersey, would help provide a seemingly more objective measure. Goddard was one of the first to bring Binet's tests to the United States, which he had seen while in Europe. Although at first he was skeptical, Goddard used the tests on the children of the Vineland School and found that it confirmed his intuitions about most children. Goddard was convinced that feeblemindedness was inherited from a single recessive gene. Persuaded of the dangers of feeblemindedness, and concerned that immigrants were carrying this trait in high numbers, Goddard profoundly transformed the interpretation of the intelligence test that he was experimenting with at his school. In Goddard's hands it was no longer the starting point for compensatory education but now revealed built-in and unmodifiable intelligence. To Binet, intelligence had many facets; to Goddard it was a single number. All talents were rooted in one general kind of intelligence, and the intelligence test measured it. In the United States, intelligence would become as fixed and measurable as height.

And so in 1911, the United States Health Services invited Henry Goddard to Ellis Island, the country's major immigration center, to administer the intelligence test and help them screen out the feebleminded. Goddard's results at Ellis Island shocked even him: 83 percent of Jews, 80 percent of Hungarians, 79 percent of Italians, and 87 percent of Russians were morons, a word Goddard dubbed to refer to those adults with a mental age below twelve years on the Binet scale. At a later time such results would be interpreted as evidence for language and cultural bias in the test, but at the time Goddard wrote with pride, "the number of aliens deported because of feeblemindedness . . . increased approximately 350 percent in 1913 and 570 percent in 1914."[5] The eugenicists' campaign would also result in the passing of sterilization laws in thirty states that would see more than forty-five thousand Americans involuntarily sterilized.

Major Yerkes's Private Ambition

Although the use of mental testing was spreading under Goddard's program, it would take World War I for the administration of the intelligence test to reach massive proportions. That accomplishment would fall on Robert Yerkes, a Harvard psychology professor. A man of high academic reputation, Yerkes's great ambition was to put psychology on as rigorous a footing as physics. In the great numbers of U.S. Army recruits, a veritable gold mine of potential data, he saw his chance.

Yerkes began a campaign to convince the army that it needed to test all 1.75 million recruits for "native intellectual ability." This was a radical idea—nothing near its scale had ever been tried before. Yerkes, however, was a skilled campaigner, and so won the army's approval. As America went to war in 1917, Yerkes arrived at Goddard's school to write the Army Alpha and Beta tests. Joining them there was Lewis Terman, the man who coined the term *IQ* and would later revise Binet's test to produce the Stanford-Binet test, whose fourth revision remains one of the standard IQ tests today. Together, they quickly devised a series of intelligence tests for army recruits. Despite substantial problems with their testing, Yerkes began an aggressive marketing campaign, transforming the public's perception of mental testing and eventually converting the arcane art of mental testing into a multimillion-dollar industry.

The eugenicists' intense public campaign and political lobbying were rewarded in 1924 with the passage of a landmark piece of legislation, the Immigration Restriction Act, which characterized national immigration policy until 1968. Also known as the National Origins Act, it limited total annual immigration to 150,000 by restricting immigration from each country in proportion to the number of residents from that country living in the United States in 1890. Since the new wave of immigration from southern and eastern Europe came mainly after 1890, it effectively shut down immigration from these countries. It also all but barred Asian immigration. On the other hand, "Nordic" immigrants had made up the bulk of the country before 1890, and so more than 105,000 yearly immigrants were allowed from Great Britain, Ireland, and Germany, a rarely filled quota.

The eugenics battle over immigration helped solidify the modern view of intelligence as a genetically transmitted trait, and the use of the intelligence test in that battle suggested a simple means of measuring that trait. The foundations of the modern view of intelligence were in place, poised to collide with other social changes in the United States that would cement this view in the public consciousness.

A New Way of Work: The First Pillar of Intelligence

Perhaps the most profound social change that was taking place in America was a new way of making things: mass production. American industrialists such as Andrew Carnegie had used mass production in the late nineteenth century, but it was Henry Ford who became synonymous with it. Inspired by the assembly-line operations he witnessed at a Chicago slaughterhouse, Ford standardized his car's design, allowing for mass production's competitive advantage: economies of scale.[6] Ford's new factory captured the public imagination as a gleaming example of twentieth-century possibilities. For a public already entranced by the image of the machine, the melding of worker and machine on the assembly line represented a new ideal that would sweep the nation—the cult of efficiency. Efficiency would find its champion and its greatest promoter in Frederick Taylor, who perfected and codified its operation through his study of workplace efficiency.[7] Taylor was the architect of "scientific management," a hugely influential approach to organizing work and public institutions in twentieth-century America—and the world. In the 1970s, American economic historians voted Taylor the most influential thinker in business history.

The key to Taylor's efficient production was supremely mechanical, involving rote, unthinking repetition. Under Taylor, the craftsman became an operator, deskilled and as replaceable as any other part on the assembly floor. One of the deepest legacies of Taylor's method was the cult of specialization: The worker no longer understood the whole process he toiled in. He was no longer making a car or a shoe. He was making a tiny piece, a part of the door assembly or an insole. As the cult of specialization became the key to efficient production, intelligence in a worker became defined as little more than the capacity to follow orders, instructions someone else had devised for a process that the worker needn't understand. The mass-production operator became the century's icon of minimal intelligence, just enough to function within a highly specialized job description.

The New Schooling:
The Second Pillar of Intelligence

For a society transfixed by the idea of rational, scientific rules and procedures in every institutional decision, the intelligence test would become the standard operating procedure for filling the differentiated slots of intelligence demanded within the new schooling as well as in industrial America.

As the twentieth century neared, public educators had reason to be optimistic. Free public schools from kindergarten through university were closing in on equal educational opportunity (though still limited to white America). In fact, the rise of public education was spreading so quickly that by 1892 there was a strong need for standards for high school curricula and college admissions. The task fell to Harvard president Charles Eliot and his Committee of Ten, an elite panel of university presidents and professors. They recommended that all children take a minimum high school curriculum of four years of English, four of a foreign language, and three of history, mathematics, and science, whether they were college bound or not.

Ten years into the twentieth century, however, Eliot's optimistic vision was in shambles. Public confidence in schools had eroded to the point where the educational system was considered an utter failure. Muckraker journalists, writing in wildly popular magazines like *The Saturday Evening Post, Ladies' Home Journal,* and *McClure's,* exposed the inefficiencies of what they saw as outdated educational aims. The March 5, 1912, issue of the *Saturday Evening Post* published "Our Medieval High Schools—Shall We Educate Children for the Twelfth or Twentieth Century?" It was a vehement attack on a "gentleman's education," the kind of education for its own sake that Eliot thought was the sign of a civilized society, but which to the muckrakers had no use in the business world. A week later another article, "Medieval Methods for Modern Children," attacked the "inefficiency in the business management of many schools such as would not be tolerated in the world of offices and shops."[8]

Ladies' Home Journal stepped up the attack. In "Is the Public School a Failure? It Is: The Most Momentous Failure in Our American Life Today," the author, Ella Frances Lynch, a former teacher, asks, "Can you imagine a more grossly stupid, a more genuinely asinine system tenaciously persisted in to the fearful detriment of over seventeen million children. . . . Can you wonder that we have so many inefficient men and women. . . ."

What was at the root of this crisis and public mistrust? The most obvious challenge confronting schools was the struggle to accommodate the wave of immigration that was so dramatically altering the American landscape. Between 1880 and 1918, student enrollment increased by more than 700 percent, from about 200,000 to over 1.5 million. To cope with such massive increases, between 1890 and 1918, on average more than one new high school was built every day of the year.

It is little wonder that a system designed for a homogeneous English-speaking population and built around the village model of the school and its administration couldn't cope with these sweeping changes. The rush into the cities meant new concentrations of urban youth, placing an incredible strain on city administrators to deal with the new educational demand. Poorly trained teachers faced large, diverse classrooms whose walls now echoed with many languages. The strain was too much for the old "village" system. Schooling had reached its breaking point.

The response to the crisis in schooling shaped the American educational system into a form surviving to this day—and with it the modern image of intelligence. Franklin Bobbitt, an instructor at the University of Chicago, now enters the story. In 1913, Bobbitt, a publicly enthusiastic supporter of eugenics, offered what historian Raymond Callahan calls the "great panacea" to beleaguered school administrators. In *The Supervision of City Schools,* Bobbitt presented the most detailed plan to date for putting schooling on a scientific foundation. Not surprisingly, his inspiration was Frederick Taylor. The appeal to science proved irresistible to school administrators.

Just as the key to Ford's success was standardization, Bobbitt felt the first great step forward was in standardizing education's product: children. To him, schools could standardize their products as precisely as industry; in his own words, "The ability to add at a speed of 65 combinations per minute, with an accuracy of 94 per cent is as definite a specification as can be set up for any aspect of the work of the steel plant."[9]

Charles Eliot had maintained that the goal of education was to train the mind for participation in a democracy, an American line of thinking that goes back to Thomas Jefferson. To administrative progressives like Bobbitt and to the muckrakers, education should make students socially efficient by preparing them to take their place in mass-production society. The influx of captains of industry into school boards helped spur the move to make education an arm of industry and to reflect its needs.

For example, the influential educator Ellwood Cubberley now saw schools as "factories in which the raw materials [children] are to be shaped and fashioned into products to meet the various demands of life."[10] Smaller schools were closed and their populations absorbed in gigantic regional schools, bringing economies of scale that reflected earlier changes in the auto industry. Even the school day, as in the factory, was divided into uniform work periods, a bell signaling their beginning and end, a regimen that continues to this day.

IQ and the Great Panacea

This rethinking of education's role in a mass-production society had a profound effect on theories of intelligence. Society's needs were diverse, requiring an increasing number of specialized occupational niches, from the foot soldiers to the general. One problem remained. Schools needed some way to choose who would be society's foot soldiers, the front-line workers, and who would be the captains and the generals. For this, we turn to Lewis Terman, who had called the intel-ligence test the "beacon light of the eugenics movement."

Whereas Charles Eliot believed every child should take the same subjects and wrote optimistically about the intellectual possibilities of the great majority of children, Terman believed that more than 60 per-cent of students were incapable of intellectual work. The intelligence test would be used to filter and assign student placement within the intelligence hierarchy. As Cubberley put it, "Increasing specializa-tion . . . has divided the people into dozens of more or less clearly de-fined classes." He argued that schools should begin "a specialization of educational effort . . . in an attempt to adapt the school to the needs of these many classes."[11]

If the intelligence test proved scientifically that the majority of children were doomed to failure, not through any fault of the schools but because of bad genes, so much the better for school administra-tors. In fact, some, like principal William Dooley, were quick to raise Terman's pessimistic estimate of 60 percent incapable to 85 percent. The useful test not only categorized and filtered but also absolved. As education historian David Tyack observes, one of the main selling points of IQ tests was that they helped teachers convince angry par-ents that their children's failures were not the fault of the schools but rather reflected the objective numbers of their IQs.[12]

For these reasons, IQ testing swept the nation as an educational crusade, with Terman its greatest promoter and practitioner.[13] In one of the most comprehensive testing programs, Terman and his former stu-dent Virgil Dickson, director of research for Oakland, California, tested thirty thousand students. Dickson states their goal with alarm-ing candor: "Find the mental ability of the pupil and place him where he belongs," which for them was one of five ability tracks.[14] His test re-sults fell along racial and social lines: Schools in affluent neighbor-hoods placed more than half their students in fast classes and only 3 percent in a limited class, while poor schools placed more than half their students in slow classes.

If there are any doubts about the link between IQ and vocation, lis-ten to what Lewis Terman has to say:

[P]reliminary investigations indicate that an IQ below 70 rarely per-
mits anything better than unskilled labor; that the range from 70 to 80
is pre-eminently that of semi-skilled labor, from 80 to 100 that of the
skilled or ordinary clerical labor, from 100 to 110 or 115 that of the
semi-professional pursuits; and that above all these are the grades of
intelligence which permit one to enter the professions or the larger
fields of business . . . this information will be a great value in planning
the education of a particular child and also in planning the differentiated curriculum here recommended.[15]

Tracking the Hierarchy

Under the intelligence hierarchy in Taylor's industrial America, the
U.S. standard of living became the envy of the world. By the mid-
1950s, nearly half of all American families belonged to the fast-
expanding middle class. Their ranks were filled mostly with skilled and
semiskilled factory workers and clerks, who went on a twenty-five-year
shopping spree to fill their new houses with the conveniences that
mass production made possible: dishwashers, washing machines, re-
frigerators, electric toasters, television sets, and, of course, cars.

The "American Method" spread throughout the postwar world, making the twenty-five-year period after World War II the greatest period of
economic growth in human history, with the world's gross product growing from $300 billion to $2 trillion. This golden age of the intelligence
hierarchy was a triumph of minimal abilities, a way of making life and
work comfortable without possessing extensive skills. When politicians
today call for a return to the educational principles of the fifties, they
forget that only 15 percent of students went on to college. More than
half didn't even finish high school. But the consequences of dropping
out were insignificant by today's standards. A dropout could readily find
an unskilled job that came with a career path, the implicit guarantee of
lifetime employment, a decent standard of living, and by the mid-fifties
health insurance, a pension, paid vacations, and soon automatically adjusted wages according to the cost of living.

Like all golden ages (and one whose racial segregation badly taints it), it couldn't last forever. In the 1970s, the shift began from a manufacturing economy to one based on knowledge: The modern view of intelligence and its ability hierarchy became an increasing threat to prosperity. Income inequality—the gap between those at the top, middle, and bottom of the income scale—has grown significantly throughout the past two decades and remains higher than at any other time in the postwar era.[16] This inequality reflects in part the redistribution of rewards brought on by knowledge's increasing value—the ability hierarchy of the modern view of intelligence does not match today's demands. Indeed, if the modern idea is that a single number can represent your inborn and inherited mental horsepower, and if this intelligence is distributed according to the ability hierarchy, then an emerging knowledge society portends major dangers. In particular, it suggests that our growing economic stratification may be the result of the population's unavoidable biological inability to meet today's economic demands. Reflecting on this, the authors of *The Bell Curve* suggest that

> over the next few decades, it will become broadly accepted by the cognitive elite that the people we now refer to as the underclass are in that condition through no fault of their own but because of inherent shortcomings about which little can be done.[17]

If we are creating a society in which many citizens are deemed incapable of productive lives through limitations of their biological endowment, then, according to these authors, the only solution is a "high-tech and more lavish version of the Indian reservation for some substantial minority of the nation's population."[18]

Alfred Binet himself warned against the "brutal pessimism" of unmodifiable IQ. Indeed, critics have vilified IQ's proponents and shot back with ideological arrows. For example, many researchers denied that IQ had any heritability,[19] suggesting instead that differences in IQ were the sole product of environmental differences. Dissenters had to take this position because both sides accepted the premise that any identification of intelligence with biology would lead to a grimly pessimistic determinism. But, as we have documented throughout this book, this premise is false. There is no deep reason to be suspicious of examining the biological foundations of intelligence—so long as you bear in mind that intelligence, like other complex mental abilities, takes shape through a rich interplay between brain and world.

Ironically, the most compelling evidence that different environments can have large effects on intelligence comes from within the field of IQ itself, in a phenomenon called "the Flynn effect." A political scientist at the University of Otago in New Zealand, James Flynn has carefully documented the steady increase in scores ever since the IQ test was invented. Among the twenty countries Flynn has studied, IQ has been increasing anywhere from five to twenty-five points in each generation. Although the exact cause of the Flynn effect remains a mystery, it is clear that a genetic explanation isn't correct—environmental factors are at work.[20]

The Flynn effect strongly suggests that IQ is not a fixed trait. Since IQ is not fixed between generations, it is doubtful that it is fixed in individuals, either. If intelligence reflects our biology's capacity for responsiveness to our environment, then a new science is required to understand the nature of intelligence, one that places the interplay between our biology and the world at its center. Cultural biology affords both a deeper understanding of intelligence and important lessons for how to begin constructing environments that maximize the brain's responsiveness. Ultimately, we believe this new understanding will allow us to create cultures of intelligence in which the overwhelming majority of people are capable of productive lives in today's—and tomorrow's—highly demanding world.

As we began thinking about the implications of cultural biology for intelligence, we realized that there was another fundamental problem with IQ that is reminiscent of a scene from the surprisingly successful movie *My Dinner with Andre*. In the film, two old, rather eccentric friends talk over dinner. Over dessert, Andre tells a story about visiting his sick mother in the hospital. She had slipped and broken her hip, but in the hospital she caught pneumonia and was now near death.

While Andre was visiting one morning, the orthopedic surgeon stopped by to check in on her and was delighted to see that her hip was healing nicely. On the way out, the surgeon remarked that she was doing wonderfully and should be up in no time.

We think the preoccupation with one facet of human intelligence, the kind that an intelligence test measures, is a lot like the surgeon who mistook the hip for the whole patient. We've noted that a tangled web of social, emotional, intellectual, and motivational systems inside your brain makes you who you are. The fact that IQ accounts for only a modest amount of our success in life indicates that intelligence is likewise a complex blend of these strands.[21] IQ may be a piece of the puzzle, even an important one, but it's not the whole puzzle.[22]

Cultural biology takes this a step further: Not only is intelligence a complex strand of social, emotional, intellectual, and motivational brain systems, but the central role of culture in our mental life reveals that intelligence isn't just inside the head.[23] This makes the interaction between brain and world all the more crucial in shaping intelligence. Champions of IQ, however, ignored the possibility that intelligence might depend in part on your world, arguing that intelligence was supposedly something that could be measured in social isolation with a pencil and paper test. To see how grossly misleading this assumption is, consider the following.

"The number is: 310-555-3425," the tinny, automated voice on the other end of the line announced. Steve fumbled for a pen, all the while repeating the number aloud, but before he could find one, the number was gone from his mind. Exasperated, he called directory assistance again, but not before making sure he had pen and paper ready.

What Steve did looks like a typical, everyday act. But it signifies a transforming fact about the human mind. The task of briefly remembering a phone number calls on working memory, which has severe limitations. To escape these limitations, Steve used artifacts—pen and paper—to hold a thought. Artifacts, then, let you overcome the limitations of your internal memory systems by using external storage devices.

The extension of minds into the world through the use of artifacts was perhaps the last vital step in the evolution of culture that underlies the modern mind.[24] Written symbols, alphabets and number systems, are ways of using the world to hold ideas. These external symbols allow a society a capacity for systematic thinking that would be impossible otherwise, a process we have referred to earlier as progressive externalization. Indeed, these external devices are not just static devices for memory storage. We have built external devices that process information, mirroring the process of thought inside our heads, at least loosely. Consider numerical calculation. You are limited in the amount of numbers you can easily add in your head. A paper and pencil increase this ability tremendously by letting you manipulate external symbols and hold intermediate steps in the calculation. By using artifacts that themselves process symbols, such as a handheld calculator, however, you can dramatically extend the realm of thought.[25]

"Computers" once referred to people who manually calculated complex solutions to problems; for example, they were needed in warfare to compute the angle at which a howitzer should be fired. The complexity of these problems spurred the invention of electronic computers, which have had a profound impact on the realm of thought. Thus, what is mentally possible for you to accomplish depends on the tools, the intelligent artifacts, you have available. Computer simulations today let college students routinely solve problems that were beyond even the greatest mathematicians fifty years ago, and have led to the development of new areas such as computational physics and computational biology. In addition, computer-aided visualization techniques allow us to see relations in enormous data sets, and will be indispensable in understanding the information latent in the human genome, an emerging field known as bioinformatics.

As intelligent artifacts become more ubiquitous—that is, as more and more objects in your environment become information-rich—the dividing line between the thought processes inside your head and the things in your world becomes harder to demarcate. The equivalent for a computer would be RAM upgrades and new peripherals. It is extremely limiting to consider thought to reside purely inside the head.

You have become smart in part by literally extending your thoughts out into the world.[26]

As we considered how the interaction between brain and intelligent artifacts alters intelligence, we realized that it had important implications for thinking about the link between IQ and individual differences that is a cornerstone of the modern view of intelligence. Consider the following analogy.

In 1936, wearing leather shoes on a slow cinder track, runner Jesse Owens clocked 10.3 seconds for the 100 meters. In 1999, wearing superlight sprinter's shoes on a high-tech track, Maurice Greene clocked 9.79, the fastest time in history. A slim half-second separates them, the culmination of sixty-three years of training and equipment know-how. With the 100 meters, even the smallest difference is magnified quickly because it is so elemental and simplistic in form. The story is entirely different for other kinds of sports. Consider the pole vault. Between 1942 and 1960, the world record improved by only 5 centimeters. In 1963, however, the world record soared upward by 23 centimeters. Was some secret training regime or chemical performance-enhancer discovered? In fact, it was a technological innovation—the fiberglass pole. Even the individuals who excelled with the old pole couldn't compete with those who had the new, flexible pole. Technology has further revolutionized the pole vault, shooting today's record up an astounding 132 centimeters beyond the 23-centimeter increase in 1963.

Your mental life depends on interacting with intelligent artifacts, making intelligence much more akin to the pole vault than the 100-meter sprint. There's no doubt that individual differences still matter, but even these can be swamped by differences in the tools you have at your disposal, which can substantially alter your performance capacity.

That an increasingly complex environment changes the brain has been cited as one possible explanation for the Flynn effect, and this resonates with cultural biology. There is little doubt that the social environment has been steadily increasing in complexity over the years, and this may have broad effects on intelligence. In studies of the nonhuman brain, there is a long tradition of research showing that complex environments, such as cages with toys and obstacles that are re-arranged frequently, initiate a cascade of growth changes that result in more complex brain circuits, while impoverished environments result in simpler circuits.[27] An interesting twist to this research is that for the enrichment effects to work, the animal must actively explore the environment, not simply passively watch it. These brain alterations are no matter of tinkering. Many billions of synapses can be made or lost this way, a capacity that lasts to some extent across the life span.

The brain science community is just beginning to examine how more specific kinds of environmental alterations can change the structure of the nervous system to improve mental performance. Perhaps the most intriguing possibility is to build on the insight that technologies play a role in shaping intelligence. In so doing, we may construct new learning technologies that take advantage of the brain's flexibility to improve mental functioning.[28] Indeed, one such intervention, a potential model for the future, has already been designed.

Intervention Through Environmental Tailoring

Reading ability is one of the strongest predictors of educational success, and, therefore, life outcome. Children who struggle to read often find themselves in a downward spiral of diminished expectations. In fact, about 80 percent of children classified as learning disabled are classified as such on the basis of their difficulties with oral and written language. Since about 40 percent of American schoolchildren are behind in their reading ability, this represents a major risk. Approximately 8 percent of all children have significant developmental language impairment of unknown origin. This delay appears to affect only language skills, since such children score in the normal range of IQ. In recent years, attention has turned to the possibility that deficits in the speed of auditory and visual processing may be at the core of this language learning impairment. The auditory systems of some children may take longer to process brief, successive acoustic signals, on the order of a tenth of a second rather than a hundredth of a second. Since speech sounds involve changes at a rate of a hundredth of a second, it is impossible for these children to segment speech adequately. Indeed, difficulties in sound processing in six-month-old infants have been correlated with delays in language learning when those children reached two years of age. More recently, researchers have found that the ability to process visual and auditory stimuli rapidly lies along a continuum that is correlated with individual differences in literacy skills.

The fact that there are basic biological differences in the ability to process visual and auditory stimuli, leading to individual differences in literacy, could be interpreted to support a pessimistic biological determinism analogous to many of the IQ arguments. In the context of literacy, must we accept the fate that biology ordains?

In 1995, Michael Merzenich of the University of California, San Francisco, a pioneer in brain plasticity, and his colleagues were training monkeys to discriminate subtle differences in rapid sequences of sounds and touches. As the animals' skills improved, the team found that the timing of the responses of neurons in the animals' brains also changed. Other research with humans showed that practice led to a great improvement in the ability to recognize brief stimuli. This raised the possibility that the deficits in auditory processing that underlie language-learning impairment might be altered with training. In 1996, Merzenich teamed up with Paula Tallal, an expert in the brain and reading at Rutgers University, to develop a computer-based training program to improve children's auditory processing. If normal speech was too rapid for some children, then speech sounds could be modified by computer to draw out these changes. This modified speech input was put in the form of a variety of computer games for children to play, where success depended on recognizing the changes in the computer-modified sounds. The game adapts as a child improves, making the sound changes more rapid. Children worked on these exercises one hundred minutes per day, five days a week, for four to eight weeks. This intensive, daily training resulted in meaningful gains in their auditory processing rate.

Based on this success, in 1996 Merzenich and Tallal launched a company, Scientific Learning Corporation. In 1997, they launched their first product, Fast ForWord. Since then, approximately one hundred thousand children have taken the program, many of them improving their reading from one to four grade levels. Preliminary evidence indicates that the program resulted in observable brain changes underlying speech processing.

Although there is still a great deal left to discover regarding the brain basis of language, this remediation program points in a promising new direction and goes beyond the brutal pessimism Binet warned of a century ago. We think that this intervention also points to a broader lesson stemming from the fact that children who aren't reading well by the third grade often lose their enthusiasm for learning.

As children develop models of their selves, their increasingly sophisticated self-conception plays a key role in how they approach challenges. Consider, for example, the early experience of Robert Sternberg, now a professor of psychology at Yale and a leading expert on intelligence:

> When I was very young, I did poorly on IQ tests because I was test anxious. The result was that teachers had low expectations for me and I wanted to please my teachers. So I met their low expectations. They were happy and I was happy that they were happy. I've been there and I've seen it happen to lots of people I know. I got over my test anxiety and then did extremely well on tests. All of a sudden the expectations were high. To a large extent it becomes a self-fulfilling prophecy, either way.[29]

Thus, the cornerstone of IQ, that the essence of intelligence is raw, unmodifiable mental horsepower, does not stand up to scrutiny in light of contemporary brain science. We have only begun to discover how to construct learning technologies to take advantage of the brain's environmental responsiveness, and how far such technologies will go remains an open question. There are other, looming concerns regarding building on constructive intelligence: Does its insight hold promise only for immature children's brains, or does it hold lessons for intelligence across the life span?

Teaching Old Brains New Tricks

How responsive mature brains remain is especially relevant to our changing world. In the traditional workplace, the great majority of a career was spent in a stable period, lasting from around age thirty to sixty. During this period, front-line worker and manager alike applied what they had learned to well-defined, stable tasks. Today, however, lifelong learning is essential both to navigate a highly dynamic workplace and to prosper in the larger world of change.[30]

As you age, are large declines in mental abilities inevitable, making lifelong learning impossible? Starting in the 1960s, studies of IQ suggested that mental ability began declining as early as the age of twenty-five for women and thirty for males. The picture was even grimmer for retired persons. For them, rapid and profound mental decline was all but inevitable. Indeed, traditionally, retirement wasn't seen as a time for self-discovery, but as necessary to remove people from roles they could no longer perform adequately.

A clue to why this was a misleading view stems from examining the method psychologists used to arrive at it: a comparison of the average intellectual performance of different age groups.[31] There was little doubt that a group of sixty-year-olds performed below the levels of a group of twenty-year-olds, but group comparisons obscured crucial information. The critical question was whether some sixty-year-olds did just as well as the twenty-year-olds. If so, that would suggest that severe cognitive decline wasn't an inevitability. In turn, that would open the door to studying the factors underlying successful aging.

The only way to get at these finer details is by longitudinal studies, following people through their life to examine the pattern of change and to discover lifestyle factors that might account for variations in aging. In 1956, Warner Schaie pioneered this approach to aging when he began the Seattle Longitudinal Study.[32] Involving more than five thousand people, the study has now been tracking them as they age for over thirty-five years.

Schaie and his colleagues have found intriguing aging patterns that differ suggestively from the traditional view. Although many people exhibit declines, others do not. Schaie and his colleagues also found that while some mental abilities typically decline after the age of sixty, others, like verbal abilities, actually increase. Based on this longitudinal study, Schaie found that one in four eighty-year-olds, a group traditionally written off, had no decline in overall mental abilities.

These and other studies force a revision of the pessimistic view of aging. In many cases, radical mental declines are not inevitable as you age. On the contrary, a closer look at the aging process has spawned the idea of "successful aging," exhibited by people who enjoy a vibrant and active life of the mind well into their later years.

One of the most dramatic examples of successful aging comes from a study of randomly chosen university professors between the ages of thirty and seventy-one.[33] As prototypical lifelong learners, professors represent an extreme of the potential benefits of lifelong learning. The aged professors showed some decline in reaction time, which is typical, and in some aspects of working memory. But what was most striking about these older professors was that they showed no decline in higher-order abilities, the problem-solving skills that are so crucial for learning. These skills underlying mental flexibility typically decline first in the normal population, but their maintenance in professors and in others who keep intellectually engaged offers strong evidence that their decline is not inevitable.

Constructive Learning Throughout Life

The ideas of successful aging and lifelong learning are transforming how the entire aging process is viewed. The earliest study suggesting that developmental processes continue in the adult human brain was published in 1981.[34] The authors of that study were interested in what happened in dendrites in humans after the age of fifty. Did they wither away, as the traditional aging view might suggest? For an answer, the authors compared fifty-year-old dendrites to those about eighty years old. The eighty-year-old dendrites were a full 35 percent *more* complex: Their total length was greater, as was their number of branches.

In an interesting twist, the authors added another comparison group: people who had succumbed to dementia. In these cases, they found that dendrites failed to expand beyond the fifty-year-old extent. In fact, these dendrites had actually become smaller. Much research remains to fully understand the differences between successful aging and dementia, but one intriguing place to look is in the work of David Snowdon, director of the Nun Study, a longitudinal study of aging and Alzheimer's disease and the largest human brain-donor program in history. The donors are an unlikely group: about seven hundred nuns of the School of Sisters of Notre Dame. Born between 1886 and 1916, the nuns are participating in annual assessments of their mental and physical life and donate their brains for study upon their death.[35]

Nuns who are more educated and perform stimulating work, such as teaching, tend to age more successfully than those who have less education and perform mundane tasks. They also live an average of four years longer. Since the nuns don't differ much in economic status, reproductive history, or—dare we say it—bad habits, we found this striking. Uncontaminated by the usual confounding factors, the Nun Study points strongly to the importance of a brain rich in connections for both the quality and the quantity of life.

The finding that senile-dementia brains do not continue to grow and the Nun Study together suggest that normally aging brains continue to form new connections through constructive learning. This insight has important consequences for our aging nation and our understanding of Alzheimer's disease, the most common cause of dementia among older people. Marked by progressive, irreversible declines in memory, language and communication skills, abstract thinking, and the ability to learn, Alzheimer's is a devastating disease, both for those afflicted and for those caring for them. Costing some $85 billion annually in health care, it afflicts around four million Americans.[36]

Although there isn't yet a cure for Alzheimer's disease, that doesn't mean there is nothing you can do to help prevent it. More and more evidence is accumulating to show that an active life of the mind can be the first line of defense against Alzheimer's. Building up more complex brain structures through constructive learning gives your brain more ammunition to ward off the onset of Alzheimer's disease. Consider that people with less than an eighth-grade education have twice the risk for Alzheimer's than those who went beyond the eighth grade. Those with both low education and mentally unstimulating jobs run three times the risk of contracting Alzheimer's than those with higher education and stimulating work.

Even more surprising is the recent discovery by Fred Gage and his colleagues at the Salk Institute. Contrary to a dogma in neuroscience, they discovered that new neurons are born in the hippocampus of adult rats, a brain region important for long-term memory. Even more surprising was their finding that more of these neurons survived in rats who were exposed to enriched environments. These rats also performed better on memory tasks that call on the hippocampus than did rats kept in an impoverished environment. So, active learning can not only make existing neurons more capable but also put newly created neurons to good use. Newly formed neurons have also been detected in the hippocampus of humans.

The new understanding of your brain's capacity for lifelong flexibility teams up powerfully with the recent results of studies of successful aging. Together, they cast doubt on the pessimistic view of inevitable mental decline and give hope to an aging nation in a rapidly changing society.

The First Element: Stimulating the Brain

The idea that the brain needs stimulation strikes us as common sense. But what isn't so well known is how profound the effects of mental stimulation can be. Differences in the amount of stimulation an environment affords can result in billions of extra neural connections. Consider the fascinating results of a study by Robert Jacobs, a UCLA neurobiologist. Jacobs was interested in seeing what effects different environments might have on the dendrites in Wernicke's area, a brain region involved in language. Looking for these effects in humans means finding natural equivalents to the enriched environments made in laboratory conditions for animals. Typically, education level is one such indicator; interviews with family members regarding a person's work, lifestyle, hobbies, and other interests can help create a portrait of how complex someone's environment was. Building these portraits and then measuring the complexity of dendrites in Wernicke's area in people who died from causes that would have little direct effect on the brain led to a striking result. The dendrites of those who lived in a complex environment—had a college degree, a stimulating job, and were mentally active throughout their life—were a full 40 percent more complex than those of high school dropouts. They were also more complex than those of university graduates who had not been mentally active after college.

Forty percent is a striking difference. If these kinds of changes are occurring in other parts of the brain, which they likely are, then environmental differences can underlie enormous differences in human brains. When similar robust results are found in other areas of health—such as the link between smoking and cancer or obesity and heart disease—prudent public health policy follows. Since complex, stimulating environments have such enormous effects on brain complexity, which in turn helps ensure successful aging and reduces the risk of Alzheimer's disease and other dementias, we should pay more attention to the design of our environments for public mental health reasons.

It turns out that it is never too late to engage the brain in a stimulating environment. Sherry Willis and Warner Schaie demonstrated in the Seattle Longitudinal Study that reengagement in later life can boost mental performance even in people whose mental performance had already significantly declined. They studied a group of two hundred people over the age of sixty-five, half of whom had significantly declined in mental performance over the previous fourteen years. Through training, most improved significantly. Among those who had declined, a full 40 percent were able to regain their performance levels of fourteen years earlier. Seven years later, they were still significantly ahead of those with similar levels seven years earlier.

This is the last link in a chain of evidence that we have built

throughout this book: From the earliest interventions that help establish the crucial bonds between caregiver and child to educational programs designed for people in their later years, creating rich,
compensating environments significantly impacts the brain and mental performance throughout life. And similarly, immersion in complex,
stimulating environments—cultures of intelligence—is an important
component of successful aging.

The Second Element: Novelty

The brain's lifelong learning capacity probably evolved to deal with an
uncertain and changing world. John Allman of Caltech has shown that
brain size in different primates is highly correlated with life span, suggesting that a longer-lived animal requires more neural machinery to
cope with the amount of change likely in its lifetime. This indicates
that complex environments must also feature novelty. As we noted earlier, experiments with animals demonstrate that the largest brain
changes occur when novelty is maintained by regularly rearranging and
changing the objects in the animal's environment, keeping the animals
in a constant state of learning.

Ironically, novelty is surrounded by powerful myths. Is it true that
no one likes change, or that you can't teach an old dog new tricks?
Consider what happens when a brain is confronted with novelty. Far
from causing a negative reaction, exposure to novelty typically activates the dopaminergic pathways that underlie rewarding effects.
Novelty has other effects, too. One of the first looks at the effects of
novelty inside a human brain was performed in 1992 using an imaging technique known as positron-emission tomographic (PET) scanning.[37] The researchers used PET scanning to look at the brain at
work as it learns a new task, in this case how to play the computer
game Tetris, first discussed in chapter 8. Because Tetris play is such
an intense task, someone first learning the game is completely absorbed in it, all their energy focused on its demands. The activity levels inside the brain of someone learning the game reveal a similar

pattern of absorption: Large areas of the brain become activated. After the person has become skilled in the game, though, less of the brain activates during the game, particularly in the cerebral cortex. It appears that the brains of accomplished players have devoted special circuits to the game, dedicated Tetris centers that have developed efficient game strategies.

The experienced players also exhibit other behavioral changes. Since the game is no longer demanding all their attention, the expert players seem to go on autopilot, playing the game while carrying on a conversation, switching from conscious to unconscious game play. Skilled players also show less interest in the game. Indeed, computer game companies recognize this and typically build in increasingly difficult levels to keep the player challenged, revealing the power of novelty to keep the mind engaged.

Even the most stimulating environment, then, can eventually require little more than a reflex response in the absence of new challenges. Novelty, on the other hand, calls on knowledge distributed widely across the brain, often recruiting parts to be used in new combinations. And it requires you to focus your attention, which brings your brain's resources to the problem at hand. Encountering novelty thus initiates a sweeping cascade of brain events, causing a flurry of activity that could result in new connections between concurrently active neurons.

The risks of avoiding novelty are real. In fact, Schaie found that people who at midlife were fixed in their ways, either by not accepting new challenges or by refusing to budge their core beliefs, were at risk for cognitive decline in later life. Schaie even suggests that such people would be well advised to take advantage of psychological therapeutic interventions to become more flexible.[38]

As we'd suspect from the close relationship between thought and emotion, flexibility and adaptability are also crucial for emotional well-being in later life. Renee Solomon and Monte Peterson, two experts in social interventions for older adults, suggest that "flexibility and adaptability are critical personality dimensions of successful aging. . . . Older persons who lack optimism, humor, and relatedness find it difficult to be flexible and adaptable and may well be at risk for emotional collapse."

The Third Element: Family and Social Context

In chapter 7, we described the health-promoting effects of human ties. Maintaining a rich family and social context is also an important component of maintaining a vibrant mental life. The Seattle study found that people who had intact families and a rich social network showed less cognitive decline. Perhaps this is a cautionary lesson for retirement. The decline in mental ability that often follows retirement may come about in part by disrupting the familiar social patterns of the workplace. Many years of interaction with coworkers and the skilled practice of patterns of behavior are suddenly replaced by relative isolation and disuse, putting the retiree at double risk.

The Fourth Element: A Positive Self-Model

One of the deeply confounding factors in explaining human behavior lies in the fact that we do not simply respond to objective features in our environment but to how we interpret that environment. For example, a face that you interpret as menacing might look playful to me, resulting in very different behaviors. So, too, the way you interpret your experiences in light of how you see yourself plays an important role in your behavior. The way you explain success or failure goes a long way toward creating or stunting your own intelligence. If you view a failure as indicative of an unchangeable feature of yourself—"I'm no good at that"—chances are you will be less likely to take on similar challenges in the future. As we age, do our changing beliefs in our ability and efficacy play a role in cognitive decline?

Looking at groups of people tracked by the MacArthur Studies of Successful Aging, UCLA professor of geriatrics Teresa Seeman and her colleagues found that a positive self-model plays a significant role in successful aging. People who feel incapable of performing at high levels shrink from hard challenges, expend less effort, and show less perseverance at the challenges they do take on. In other words, they become at risk for the pattern of disuse that underlies cognitive decline.

The risks of retirement are clearly illustrated by a study that compared two groups of seventy-year-old doctors: One group had been in retirement for a decade, while the other was still practicing. Not only did the still-practicing group show less mental decline, they felt more capable. They also felt more useful and efficacious because other people depended on them. Not least of all, they remained embedded in a rich web of social relations, which retirement often severs. Our brains seem poorly suited to retirement. Indeed, the traditional notion of retirement remains a vestige of the industrial age that is doubly out of step with our times, in which life expectancy is soaring. The perseverance of the myth of inevitable mental decline, however, still fuels many older people's feelings of diminishing self-efficacy, as well as many discussions regarding the lifestyle and care of the elderly. Overturning these myths and providing opportunities for lifelong learning and significant engagement is a crucial challenge for an emerging knowledge society.

The Fifth Element: Able Body, Able Brain

The division between mind and body runs so deep in our culture that we tend to think that training the mind and training the body are two different things. There's no doubt that many highly intelligent people fall short on athleticism, and indeed often attempt to enhance their cerebral image by downplaying their physical abilities. Einstein, for example, became such an icon, despite the fact that he was an avid basketball player during the time he formulated his theory of special relativity. Our society also typically represents the physically fit as intellectually vacuous, as though you can only possess one ability or the other. Is this separation, too, a myth?

Many studies reveal a direct link between exercise and intelligence. College students who undertook an exercise program, for example, found that their academic performance increased. Such results are not just limited to the young. In fact, some of the most dramatic results come from studies of older people. Fifty-year-olds who were put on a four-month walking program improved their mental performance by 10 percent over their prewalking levels.[39] The MacArthur Studies of Successful Aging found a similar increase in a group over the age of sixty-five. Exercise can not only make you smarter but also make you smarter longer.

William Greenough and his colleagues at the University of Illinois are finding some intriguing clues from brain science to support this link.[40] The brains of rats that had been placed on an exercise program grew more capillaries, carrying more oxygen-rich blood to brain cells. The brain is a greedy consumer of oxygen: Aerobic metabolism is its primary means of energy. Aerobic conditioning in muscles also results in increased numbers of capillaries, suggesting a striking analogy between mental and physical fitness. Just as a muscle is able to do more work through the physical changes brought about by aerobic conditioning, so too the brain can do more work when in good aerobic shape, a finding confirmed by improvements in mental concentration with aerobic conditioning.

Physical conditioning results in the growth of new muscle cells. It turns out that exercise also has this analogous effect in the brain. We've mentioned that new neurons are continually being produced in your hippocampus, a discovery made by Fred Gage at the Salk Institute that overturns an age-old dogma. Terry collaborated with Gage to examine whether exercise had any effect on these new neurons. They found that when adult mice use an exercise wheel, more of the new neurons survive and the mice learn better than do sedentary mice. A form of synaptic strengthening that occurs when inputs to the hippocampus are stimulated is also stronger in mice that exercise, evidence that these new neurons may provide the hippocampus with the potential for making new memories.[41] Along with well-founded benefits to the body, these additional findings relating mental performance to physical fitness are yet another incentive for starting an exercise program consisting of at least thirty minutes of aerobic exercise three to four times per week.

On reflection, the link between physical and mental fitness shouldn't be so surprising. Our brains evolved to engage the world and to navigate the challenges of complex physical and social environments, not to sit around passively. These challenges don't just tax your muscles—they tax your brain to make decisions quickly. Many forms of exercise tap into all these demands. Basketball, for example, requires high vigilance, navigating around opponents, deciding instantaneously whether to shoot or pass off, calculating where a rebound will likely fall, all in a confined space full of shouting, distracting players, many of whom are doing their best to thwart your efforts. Yet our culture continues to separate mental and physical performance. Indeed, the influential theory of multiple intelligences of Harvard's Howard Gardner distinguishes between kinesthetic, or bodily intelligence, and other forms of mental intelligence. These are largely false divisions: Brain functions, including motor control, are highly integrated and crosscut these "multiple" intelligences. We believe that progress in understanding the neural basis of intelligence will require abandoning these divisions, as indeed there is little empirical support for distinct, multiple intelligences.[42]

Not only does exercise improve mental performance, it also appears to fight off mental declines later in life. In a study on rats, Carl Cotman and his colleagues at the University of California at Irvine found that regular exercise increases the gene expression of neurotrophins, a family of transmitters that support brain cell growth, function, and survival.

This finding is all the more intriguing when you consider where this increased neurotrophin expression was found: in the hippocampus, a brain region crucial for long-term memory. Damage to the hippocampus is thought to be an early event in Alzheimer's, and hippocampal shrinkage distinguishes Alzheimer's patients from age-matched control subjects.[43] Researchers can even detect hippocampal shrinking in subjects with mild age-related memory impairments—perhaps the first warning signs of grimmer losses to come.[44] If imaging techniques could be used to measure the size of brain structures in middle-aged people to determine if someone is at risk for Alzheimer's disease or other problems, it could be the first step in starting intervention programs to ward off the ravages of Alzheimer's.

The link between physical exercise and intelligence isn't seen just in old age. The biggest gains can be found among children, which isn't too surprising since their brains are most responsive to changing demands. A study of five hundred schoolchildren found that those who spent an hour each day in gym class performed better on tests than those who were inactive, a finding that has been replicated in dozens of studies.[45]

Sadly, our schools and society at large seem to be paying little attention to those results. According to the President's Council on Physical Fitness, only about a third of American schoolchildren participate in daily physical education. More than half of all children don't get enough exercise to develop healthy hearts and lungs. This inactivity not only puts a new generation at risk for physical problems later in life but also stunts their full intellectual development and puts them at risk for later mental decline as well. Again, the cultural myth separating physical and mental fitness obscures the formation of sound social policy, this time affecting the mental and physical health of the young.

Supporting Lifelong Learning

Although it is difficult to predict the direction of social changes, there are many hopeful signs for creating cultures of intelligence, the complex and engaging environments required for our lifelong mental and physical health. The emerging knowledge society—with its emphasis on complex work environments and lifelong learning—promises to bring the demands of life and work more into alignment with the needs of the brain. As such, it is undoing many of the artificial limitations the forces of the twentieth century imposed on the brain and its most remarkable property, our unparalleled intelligence. Lifelong learning lasting well into your later years isn't simply wishful thinking. The new understanding of the brain reveals that it can continue to function at high levels for a lifetime. A closer look at the aging process also reveals models of successful aging.

Dismantling myths about the aging brain is a first step toward renewing the brain's power across the life span. However, the demands of the knowledge society require that we support and foster not just occasional retraining efforts but a lifetime approach to continual learning. The intertwined threads between lifelong learning and mental, emotional, and physical well-being suggest that a cultural commitment to lifelong learning will bring with it an entire spectrum of benefits as yet untapped and only beginning to be known.

The Search 10 for Happiness

What has made national boundaries obsolete in so much of Eastern Europe, Africa and Asia? Not the Internet but tribes. What have the breathtaking advances in communications technology done for the human mind? Beats me.

—*Tom Wolfe*, Hooking Up

Sitting next to Diane Sawyer on the set of *Good Morning America*, Terry felt right at home.[1] The set of the popular early-morning TV show was designed as an extension of the viewer's living room, including the country curtains. Diane wanted to know whether it was possible to defeat the computer system that Terry's lab had developed, which could detect falsehoods by analyzing facial expressions.[2] Yes, Terry replied, it's called "method acting."

The ability to "read" others is so much a part of our notion of what it means to be human that the thought of a computer being able automatically to detect insincere expressions is deeply disturbing. This leads to another thought: Is it possible to dig further into the brain, underneath facial expressions, to detect underlying thoughts and feelings? Yes, it's called "brain imaging." Still, that technology is only in its infancy. Although we can now "look" into your brain while you are reacting, remembering, thinking, and dreaming, we don't know enough yet to fully understand what we are seeing.[3]

Will it be possible someday to find our human core and begin to understand our real feelings and needs? We saw in chapter 5 that emotions are our first response to crucial life events. Human emotions are involved in everything from sexual attraction, playfulness, curiosity, and pleasure seeking to vigilance, defense, and attack.[4] Emotions evolved to facilitate rapid problem solving and your response to complex situations where rational thinking alone would be excessively slow and inefficient. Emotions such as fear and anger focus the mind and make you pay urgent attention, priming you for action. There are large individual differences in emotional expression, as John Gabrielli and his colleagues at Stanford have begun to explore with brain imaging. These differences are partly due to genetically based variations, including the speed and strength of an emotional response, the ability to control a response or the time needed to recover from it, and the malleability of emotions. Despite much variability, there is also a common human expressive vocabulary: Happiness, sadness, fear, anger, surprise, and disgust have the same facial expressions in all cultures.[5] It is in this shared emotional human core that we begin our search for the source of happiness.[6]

More than two millennia ago, Aristotle suggested that the goal of life is *eudaimonia,* a complex idea that is traditionally translated as happiness. Something essential, however, gets lost in translation, for Aristotle meant something more akin to human flourishing—the good life. Aristotle maintained *eudaimonia* could be achieved though civic engagement that fosters uniquely human capacities. For more than two thousand years, moral philosophers debated what constitutes the good life, but few denied either that it was attainable or that it was the proper goal of life. The rise of Darwinism changed all that. It is hard to make sense of the very notion of a good life in the light of evolution. Why should evolution care about your happiness? If you are essentially a reproduction machine designed with no final goal in mind other than to pass on your genes, then your personal happiness seems at best incidental to the main business of life. Indeed, there are compelling reasons to think that the desires and goals that evolution has crafted for you are deeply incompatible with your personal happiness.

Consider the human compulsion to provide explanations for what goes on around us. This is largely adaptive, as we have seen. Yet this compulsion drives us to create explanations, to find some deeper hidden logic in life—even when there may be no real explanation. Coupled with the crucial ability to see ourselves as enduring beings with a future—a faculty central to our capacity to engage in long-term reciprocal relationships—our compulsion toward such explanations creates the need to construct a coherent life story to give our existence deeper meaning. This search for meaning and coherence is profoundly challenged by our knowledge of our own mortality. Like an existentialist trickster, evolution has instilled in us the deep need to search for meaning in a universe that ultimately may be devoid of any. Existence seems to be a problem only for humans, growing out of our capacity for reflection and our compulsion for coherence. This may be why humans have created some one hundred thousand religions over the course of history, in every society that has ever existed.[7]

Despite the contradictions to happiness that evolution appears to have built into us, evolution has also built in a remarkable resilience. Even in the bleakest situations of profound misfortune, when hope appears lost, humans nonetheless often find a way to continue on, frequently triumphantly. This again is the selective pressure of evolution at work, for those ancestors who could carry on in the face of extreme adversity were rewarded in the genetic lottery of procreation. In this light, happiness may not only be an attainable goal, but an evolutionary imperative. In this final chapter, we turn to examine the implications that cultural biology may hold for the search for happiness.

Happiness and Human Nature

The search for happiness is intimately tied to assumptions regarding human nature and the human condition. Consider, for example, the implications of many evolutionary psychologists' insistence that a universal human nature was shaped millions of years ago out on a distant savanna. According to this view, since it's unlikely that you will ever return to life as a hunter-gatherer, your prospects for happiness in the modern world appear dim: You are a Stone Age creature, inevitably out of step with the world around you. Cultural biology, in contrast, suggests that who you are is not the product of an unmodifiable human nature. Rather, your internal guidance system creates basic needs in you that interact with the cultures in which you are embedded. The human condition, then, is the ever-evolving product of a dynamic interplay between your history and your biology.

This is not to say that the human condition is simply shaped by culture, the traditional alternative to biological determinism. Attempts to re-create human nature through mass education and propaganda efforts, such as those of the Soviet Union and Communist China, failed in part because the human core doesn't adapt passively to changing contexts. It is, rather, like a reactant with its own intrinsic properties creating universal human needs. The invidious consequences of the twentieth century's mass social experiments illustrate what can happen when a society is built on principles inconsistent with the human core.

Ultimately, the human search for happiness depends on unraveling the dynamic of brain development and the influence that social life has upon it. This enterprise constitutes a nascent neurosociology, an approach to understanding the human condition and possible avenues for its betterment that recognizes the power of both our biology and our social context to shape us.[8] We want to better understand how social contexts collectively create human personality and thus help determine the prospects for our search for happiness.

The Rapid Rise of Cities

The Santa Monica Mountains rise straight out of the Pacific Ocean, their steep peaks the visible sign of great subterranean struggles. On a recent mountain bike ride through them, Steve noticed a herd of deer foraging up ahead. Although deer are a common sight there, this was an unusually large group, numbering around a dozen. As he approached the herd, he crested the peak of a sharp hill. The landscape of rugged peaks and deep canyons had such a timeless aura that he could almost picture a group of distant ancestors lying in wait for the deer. Instead, the Los Angeles basin stretched as far as the eye could see, an urban landscape of breathtaking expanse. Office towers shimmered in the late-afternoon light. Over them glinted a long procession of planes on approach to Los Angeles International Airport. The incongruity of the two scenes was jarring.

This contrast is all the more remarkable when you consider how recently urbanization reshaped the landscape. A little over a century ago, the inhabitants of the Santa Monica Mountains lived much as our distant ancestors did, as evidenced by the local cave art and stone artifacts. In the previous two chapters, we chronicled some of the torrential forces that reshaped life in the early twentieth century. These same forces reshaped the landscape. In particular, with the rise of industrialization, the village way of life disappeared as the modern rushed into every facet of human experience. It is remarkable to realize that in 1870 fewer than 20 percent of Americans lived in cities with populations over eight thousand. A mere fifty years later, half of the U.S. population were city dwellers. Today, cities have metamorphosed into metropolitan areas that claim just 16 percent of the country's land but hold 80 percent of the population. Country life has also changed: The family-oriented farms of the nineteenth century are found today mainly in picture books, having yielded to industrial-scale agriculture in the twentieth century.

As we begin a new millennium, understanding who we are requires uncovering what changes in the human condition occurred with the breathtakingly rapid rise of the modern. In particular, it requires that we look at both the modern design of life, from its institutional arrangements to the design of our living spaces, and its culture—the material from which we fashion our self-conception and formulate our prospects for happiness.

Consider first the profound shift in culture that coincided with the transformation of a small-town America into an urban one. As we explored earlier, the core of modernism was Hobbes's sovereign individual as the basic unit of society. As the German sociologist Max Weber chronicled in his great work *The Protestant Ethic and the Spirit of Capitalism,* capitalism flowed out of this conception of the individual and the Protestant ethic as the rationalization of life, making economic activity the fundamental mode of human existence. Small-town America was accordingly dominated by the Protestant ethic and the Puritan temper, which stressed sobriety, frugality, sexual restraint, and production. The rush to the cities transformed this self-conception. Eager to throw off the shackles that ruled small-town life, intellectuals flocked to Greenwich Village in New York City and other centers to create a new Bohemia, for whom freedom meant, among other things, the liberation of sensual desire. Above all else, the Freudian view of human nature shaped modern culture, as was clearly seen in the infusion of Freudian psychoanalysis into virtually every realm of human experience, from the arts to economics. It was Freud's view of sexual desire, in particular, that undermined the sovereign individual, dislodging the rational self's restraint with the willful self's ceaseless appetite for the new.

Coinciding with the emergence of an urban middle class, an economy centered on production thus became transformed into one based on consumption.[9] Rooted in the "new capitalism" of the 1920s, this was the first economy designed to feed the lifestyles, rather than the stomachs, of a mass culture, culminating in the consumerism of the 1950s.[10] As Harvard sociologist Daniel Bell notes, "culture was no longer concerned with how to save and achieve but how to spend and enjoy. Despite some continuing use of the Protestant ethic, the fact was that by the 1950s American culture had become primarily hedonistic, concerned with play, fun, display, and pleasure."[11]

A massive shift toward the urban experience. A new culture centered on the image of the consumptive self. It would be remarkably surprising if these titanic changes in the design of life and culture didn't have an impact on human happiness. One clue that such an impact was both real and deep comes from a nagging paradox of the post–World War II period of economic growth.[12] Noting some of the bestselling books of the 1950s perhaps best captures it. Consider J. D. Salinger's *The Catcher in the Rye,* which features a narrator, Holden Caulfield, who is unable to make connections with the world around him. In sociology, David Riesman's *The Lonely Crowd* broke out from the academic world to become a surprise bestseller. Indeed, the major sociological themes of the 1950s were despair, anomie, and alienation. How could an economy of unprecedented productivity that was devoted to a fulfilling lifestyle result in a vague discontent, rising levels of depression, and a growing sense of purposelessness? How could economic and mental well-being be so at odds?

One suspect is the nature of the new work itself. As cultural anthropologists have long told us, the goal of a career, from the hunter-gatherer society to preindustrial society, was to fulfill deeper needs, to achieve both physical and symbolic satisfaction in life, a place in the community, and a sense of purpose. The modern approach to work almost totally neglected how people would react to the work they performed. Worse yet, the architect of scientific management, Frederick Taylor, believed that people were passive by nature, and so designed his systems of scientific management around issues of control. It didn't take long before the dangers of neglecting the social context of work became clear. After an initial excitement over rising wages, the daily grind of the new work began taking its toll on workers. By the second decade of the twentieth century workers were already reeling from the fatigue and physiological breakdowns it caused. Industrial psychology was thus born in an attempt to understand the psychological battering the new work dished out.

With the modern way of work was thus born the company or organizational man, chronicled in William Whyte's 1956 work, *The Organization Man.*[13] Whyte bemoaned the widespread "passive ambition," mass conformity, and corporate compliance of organizational men. Some twenty-five years later, Studs Terkel's classic book *Working* revealed the stories of Americans struggling to find fulfillment in jobs that held little opportunity for meaning, stripped work of its wider significance, and didn't allow for personal autonomy. The typical double-digit rates of absenteeism for front-line workers was just one more indication that work provided little more than an instrumental means for wage earning. Poor quality, low morale, and little commitment to the process were the more subtle symptoms of an alienating work structure.

Elton Mayo, a pioneer of the human relations movement, had contributed some key insights into the social dynamics of the workplace. In a now classic study of work between 1927 and 1932 at Western Electric's Hawthorne plant in Illinois, Mayo and his colleagues presented the first detailed look at the social nature of work. They discovered that, despite the best efforts of management, workers were actually banding together in informal social groups. The norms of these informal groups, not the efficiency expert, determined how much work would be done every day. What's more, no amount of incentive pay budged workers from their shared but implicit ideas of what constituted a fair day's work.

Mayo also found what today is known as the "Hawthorne effect," that just asking workers for their input caused their productivity to rise, even if no real action was taken. By acknowledging the social nature of work, by making workers feel as if they belonged to a team, their attitudes toward work changed and they became more committed. Mayo succinctly summarized his basic belief in the primacy of social context: "Man's desire to be continuously associated in work with his fellows is a strong, if not the strongest, human characteristic. Any disregard of it by management or any ill-advised attempt to defeat this human impulse leads instantly to some form of defeat for management."[14]

Many of the changes within the workplace—hierarchical control, the lack of autonomy and ownership of processes, neglect of basic needs created by our biologically rooted sociality—spilled over into everyday life. For example, changes in the nature of work and the products it produced, particularly the automobile, had a profound effect on the physical structure of our living spaces. Whereas industrialization first moved work out of the home and into the city, after World War II a new migratory pattern emerged—the rise of the suburbs. Spurred by the largest engineering project in history—the American highway acts—the automobile and the resulting suburban sprawl shifted life from a local level to a regional one. Life's processes, once anchored around a town center, became diffused into an often inefficient and isolating design.[15] Without romanticizing small-town life, one should not underestimate the effects of this shift in the design of living. Built largely without the benefit of town planners, the suburbs reduced people's opportunities for maintaining the social contact necessary for reciprocating relationships.[16] As the suburbs spread, mobility rates surged. By the 1950s, fully 20 percent of the population moved in a given year. Although the impact of a community's physical design on public mental health is often overlooked, the contrast between the modern design of life and virtually the rest of human social life throughout history is striking.[17] Indeed, one of the cruelest things to do to one of our ape cousins is to impose social isolation.[18] Yet 26 percent of Americans live alone today, up from less than 10 percent in 1940. As science writer Robert Wright notes, it is probably no accident that the rise of feminism corresponded to the suburbanization of life. The "housewife's syndrome" that Betty Friedan wrote of was the clinical depression of social isolation, with depression rates doubling in the last twenty-five years.

Modern life, with its emphasis on work and consumption as the center of human existence and its radically new design of life, provided a social context unlike anything the biological core had ever experienced before. The result, the modern human condition, testifies to your brain's capacity to adapt to new and unexpected contexts. Yet modern life also presents troubling challenges for the prospects of human fulfillment, as the rising rate of clinical depression suggests. According to Emory University sociologist Corey Kees, roughly 20 percent of the U.S. population say they have nothing good going on in their life, a troubling sign. Antidepressants are the most prescribed drugs in the country. Is the modern design of life in some deep sense inconsistent with our human core? To search for an answer, we turn to an idea we first explored in chapter 7: social capital.

The Modern and the Civil

The human norms of trust and cooperation underlying social capital may have first emerged as a response to ecological instability. The complex social life they make possible would have been hugely advantageous to those who could commit to others for mutual advantage. The human attachment mechanisms that make social capital possible make us deeply sociable and naturally incline us to group life. So fundamental is group life to who we are that the quality of human ties is strongly linked to mortality. Indeed, the quality of social bonds is the most powerful predictor of life satisfaction.

In recent years, a tremendous amount of discussion has centered on an idea intimately related to social capital: civil society. We can divide society into three sectors: the public sector of government, the private sector of the market, and a third sector, civil society. Civil society, then, refers to associational and communal life independent of the market and the state. These are voluntary associations and informal networks: foundations, churches, public interest groups, civic groups, social movements, and the like. These groups generate what Harvard political scientist Robert Putnam calls "the 'norms of generalized reciprocity' and 'trust' that are essential components of the 'social capital' needed for effective cooperation."[19]

Alexis de Tocqueville's observations of America in the 1830s, which formed the basis of his landmark work *Democracy in America,* led him to believe that the genius of America lay in its civic involvement: Liberty was local and civic activity was prevalent; a modest governmental sphere and an unassuming private sector were overshadowed by an extensive civil society tied together by school, church, town, and voluntary association.

Today, there is an extraordinary amount of debate over whether civil society is in decline,[20] a debate Putnam sparked in 1995 with an article entitled "Bowling Alone." In his book of the same title, Putnam presents a wealth of evidence from five hundred thousand interviews over the last twenty-five years to show a marked decline in activities that build social capital: We belong to fewer organizations that hold face-to-face meetings, know our neighbors less, and socialize with our friends less often.

Is TV the Culprit?

Putnam suspects that the decline in social capital and civic society may be due to the television. Like pets, whose parasitism we discussed in chapter 7, television is a social parasite that takes advantage of brain attachment mechanisms. TV programs tend to emphasize the type of long-term reciprocal relationships that primates favor. It's probably no coincidence that the longest-running shows are soap operas, which feature both enduring communities and characters. Nor is it a coincidence that at a time when commentators write of the decline of civil society, the most popular TV shows are those that center around friendship, from *Cheers*, where everyone knows your name, to *Friends*, where everyone knows your game. Even *Seinfeld*, which was supposed to be somehow postmodern, was as traditional as *The Waltons* in terms of its core friendships. After all, how many New York City apartment doors are never locked so that friends not only drop by unexpectedly but can enter unannounced? What is striking is the depth of emotional attachments viewers form with these characters, attachments that in the end are unfulfilling because they are fabricated and not reciprocal. Thus, ironically, TV watching is spurred by sociability yet thwarts genuine social connectedness.

It is difficult to say whether TV viewing is a cause or a consequence of social fragmentation. It is most likely that TV viewing and the social changes we sketched in the preceding section feed off each other, together creating the drop in public life. Television fare is a substitute for the bonds that the modern way of life makes difficult to maintain. Although the root causes are many and complex, the structure of social life in many parts of the world today is highly impoverished,[21] hindering our natural biological tendency to seek one another's company in long-term reciprocating relationships.

Consider the political sphere. The Communist state dislocated local mechanisms of association and denied citizen ownership over decision making, reducing public involvement and transforming citizens into passive consumers of services. In the early Soviet Union, centralized decision making and heavy bureaucratization were justified by a new intelligentsia, who feared that local civic associations were a potential threat to state stability. The result was a destruction of intermediary networks, whose absence is clearly felt in Russia today. Even Freud thought the local scale was essential to human well-being; he predicted the collapse of communism because its scale was inconsistent with patterns of human behavior.

Although its causes were different, the centralization of power and the nationalization of citizenship occurred in the United States as well. The rise of industrial America resulted in an enormous concentration of economic power in the hands of industrial capitalists. As Michael Sandel notes, "Americans long accustomed to taking their bearings from small communities suddenly found themselves confronting an economy that was national in scope."[22] The political reaction to these new realities, from the New Deal to the Great Society, was an attempt to narrow the gap between social and economic life by nationalizing American political, social, and economic life. Life's stage was thus recast from the local to the national, resulting in an impoverished vision of citizenship and community. Even on the local level, the recommendations of the Taft Commission on Economy and Efficiency facilitated the replacement of village models of governance with professional local governments of bureaucrats, following sociologist Max Weber's argument that rational bureaucracy was the very essence of modern life. The participant citizens Tocqueville marveled at became passive consumers who paid their taxes in return for local services. With bureacracies built on the wrong scale and denying ownership over the po-

litical process and local decision making, the resulting lack of autonomy bred passivity and disengagement—a danger Jefferson warned of
two hundred years earlier. In this regard, the modern state frustrated
the need for human connectedness, which instead manifested itself in
massive alienation and anomie. The iron cage of modern society was a
poorhouse for a brain driven by the need for human association.

A view stemming from the Enlightenment holds that from the first
rudimentary social groups of our distant ancestors, human history has
been a march to ever more complex forms of social organization. Thus,
spurred by a capacity for both innovation and sociality, the bonds of
kinship gave way to the ties of citizenship, and the village gave way to
the town, then to the city, and, ultimately, to the nation-state. Its culmination in modernism isn't the nation-state but the global society.
And so, when most pundits consider the sorts of technologies that
were featured in the story of Rob Hall in chapter 9, they view them as
the engine of globalization.[23] Are today's technologies combining with
other forces such as the free flow of capital to propel us into a global
society? Is the global village a placeless society,[24] in which no one is attached to place? Perhaps we are becoming digital nomads, energized
by the pulsations of a global electronic nervous system.

There is a problem in the globalization vision. Consider the following few currents of change. In the first four years of the 1990s alone,
the number of independent nations more than quadrupled, unleashing
a rushing tide of ethnicity. Pundits worried about Balkanization and a
return to the social fragmentation of the Middle Ages.[25] How could
modernism end not in globalization but in premodern forms of social
life?[26] Does modernity ultimately unravel itself?

In the United States there is another telling pattern of change. Political scientist Richard Nathan calls it a bigger consequence for U.S.
domestic policy than Lyndon Johnson's vision of the Great Society. He
is referring to the massive transfer of power from Washington to the
states sparked by the 1995 "devolution revolution," which continues in
2002 with George W. Bush. Starting with the devolution of family welfare (the Aid to Families with Dependent Children program), Medic-
aid, and child welfare services, this movement from centrally

orchestrated programs to those at state and local levels is part of a larger, worldwide movement. According to the World Bank, out of the seventy-five developing and transitional countries with populations greater than five million, all but twelve claim to be embarked on some form of transfer of political power to units of local government. According to the Urban Finance Conference's recent findings on the worldwide trends in devolution, devolution is met worldwide with the reinvention of local government. Even in the former Soviet countries, where economic reform gets most of the media attention, a massive effort to establish freely elected local governments and devolve power to them has been an equally important thrust.

What are we to make of these patterns of social change? If modernism is defined by the rise of the nation-state, do devolution and the return to ethnic identity portend the revenge of the local over the global? The human core creates a bedrock truth about who we are: Although we can participate in communities built on many scales—the family, the neighborhood, the town, the nation—our biology drives our primary engagement to be with a local, enduring community.[27]

From the perspective of the brain, there is evidence suggesting that the size of primary communities was very limited indeed. The British anthropologist Robin Dunbar has examined the relationship between cortical size and group size in a number of primate species. Since complex group life requires a great deal of brain power to keep relationships straight and navigate that web of relations smoothly, brain capacity might place an upper limit on group size. Indeed, Dunbar finds a correlation between cortical size and the actual size of primate groups. Extending his findings to humans, Dunbar suggests that the size of the human brain predicts a maximal group size of around 150.[28] That's a far cry from the millions of people living in modern cities. But, then, cities have only been around for about three thousand years. Further, cities break down into districts, which break down into neighborhoods. Dunbar suggests that clans in traditional societies number about 150, and that number appears to be about the maximum number of people in cohesive social units. Experiments designed to measure the number of close friends and acquaintances that people have repeatedly turned out to be about 135. Small businesses of about 150 can operate without hierarchical management structures. Once human groups become bigger than 150, they require hierarchical management structures.

It is imperative for military groups to foster strong bonds among members, since they are literally placing their lives in each other's hands. Indeed, from the seventeenth century on, the size of the basic military group, or company, has been in the range of 150. At this level, personal contacts and loyalties can keep the unit working as a cohesive and coordinated group. One isn't fighting so much for queen or nation as for one's friends. This is also roughly the number of people that would fit inside the town meeting halls that Tocqueville marveled at in 1830s America. We guess it's probably about the number of patrons that would have fit comfortably inside Cheers.

There is another telling change in society today that might underlie the decline in social capital. As we examined in chapter 4, ecological instability propelled our ancestors toward the protective umbrella of increasingly complex group life. Much of the impetus toward a more complex form of group life was likely to build new protective layers between the individual and ecological risk. Indeed, the rise of cities intriguingly coincided with new weather patterns like, for example, El Niño.[29] In most parts of the United States, however, community members no longer participate directly in helping buffer the misfortune of other members. That is, neighbors do not depend on one another during times of hardship or misfortune, since these functions have largely been transferred to governmental agencies and services. In contrast, Steve lives in the "disaster epicenter" of the United States, Topanga Canyon, just inland from Malibu. The 1993 Malibu wildfire was just the latest major fire to ravage the area. In the past eighty years, Topanga has suffered a major wildfire—one that burns at least five hundred acres—at least once in every decade, often twice. Combined with the fear of mudslides and even earthquakes, ecological disaster is on everyone's mind (although an "irrational" place to live, its natural beauty and the refuge it offers from urban Los Angeles make it compelling). When disasters strike, there is a limited capacity for external agencies to respond. Thus, neighbors depend on one another for mutual aid in potentially life-threatening situations. It is intriguing to note that a large number of volunteer civic associations exist in Topanga, such as the Topanga Coalition for Emergency Preparedness, the Arson Watch, the Disaster Response Team, and the Equine Response Team. Another intriguing fact about the community is that it has an unusually high number of other civic associations, including a community house built and maintained by volunteers and a large number of community functions, from annual fund-raisers for the Co-op Preschool to charity fairs. Although anecdotal, it is interesting to consider whether the mutual dependence that ecological hazards create might also provide the impetus for these other civic associations that foster a cohesive community.[30] If so, then the fact that many other communities no longer serve as buffers against misfortune may mean that members lack the pragmatic impetus to foster a cohesive community.

Group Life and Happiness

All this brings us back to Aristotle, who argued that our deep-seated sociality meant that happiness required life in society. Aristotle wasn't proposing that we sacrifice our interests for the good of the state. Rather, he argued that it was in our interest, given our deeply social nature, to participate in civic life in order to fulfill ourselves. It was this understanding that Jefferson echoed when he changed John Locke's "pursuit of property" to the "pursuit of happiness." It is also where Jefferson parted company with the tradition of both Hobbes and Locke, whose asocial view of humans Jefferson called "a humiliation to human nature." Instead, Jefferson sided with Aristotle's view that we are spontaneously social, endowed with basic social dispositions for life in a community. So strong was Jefferson's belief in this view of human nature that he believed the pursuit of happiness required dividing counties into small, citizen-governed wards.

Needless to say, Jefferson's vision of happiness did not prevail. Instead, the modern conception of happiness was rooted in a Freudian vitalism, a consumptive hedonism. In the 1920s, when this new consumption ethic was linked to mass production to create the new capitalism, for the first time in history an economy was created that was not centered on production of enough goods to fulfill human needs but on consumption dedicated to fulfilling human wants. With this economic transformation, images of mass consumptive culture helped to create new wants in the public consciousness. Images touting the virtues of consumption worked in part by creating a sense of inadequacy in the consumer. Mass culture thus began to surround the consumer with images of the good life involving celebrity and success, instilling in the ordinary person a taste for exquisite comfort and sensual refinement that were out of reach for most people. Like Hobbes's individual, who searched unendingly for power after power, the consumptive self could be temporarily satiated but never satisfied.

The consumptive self thwarted the very happiness it sought by constructing a self-image predicated on inexhaustible desires. Even more insidious, the devotion to hedonism is at once a private pursuit and one that fails to make life coherent. That is, it fails to provide any meaning beyond the immediate gratification of the senses and so is inadequate to connect life to a project larger than one's sphere of private enjoyment. The result for many is a deep listlessness. As a result, we witness the spiritualization of the workplace in the self-help writings of business gurus like Stephen Covey. But that, too, is ultimately a private philosophy that fails to provide a sense of connection to anything larger than one's Rolodex.

This is, of course, supremely ironic in an age of self-improvement. The basic flaw of much self-help philosophy is its sole focus on the consumptive self. As we have stressed throughout this book, you are a complex blend of private and social selves. Who you are takes shape through being situated in a larger social enterprise that helps fulfill the needs of your public self. For a storytelling species that craves coherence, this requires a story of who you are that invokes participation in some common purpose. Perhaps the lessons of cultural biology will help spur a shift in the concept of self, a recognition of the blend of private and social selves and our inborn need for human community.

Building Engagement

The twentieth century was in many ways a long, curious experiment in seeing how far institutional dimensions could be stretched, with the notion of economies of scale as the guiding principle of the design of social arrangements, from the structure of living spaces to the workplace and schools. Although you are a member of a species that is a champion at adaptability, there is a limit to how much your brain can adapt. The disengaged brain—reflected in the loneliness and alienation of modern society, the alienating workplace,[31] and the social isolation of the elderly—is increasingly at the root of mental health maladies. Your complex blend of private and social selves and your deep-seated need for sociality suggest that engagement is a major requirement of human happiness. The lessons learned from the biological core, we believe, should be applied to daily life: more opportunities for engagement beyond the private pursuit of happiness and more meaningful engagement in a common purpose.

At the most basic level, this requires rethinking the structure of our social contexts, from how we design communities to facilitate chance meetings with neighbors to how we incorporate new technologies to foster rather than hinder social interactions. At another level, it requires devoting more attention to how we design everyday tasks and tools. One of the most rewarding experiences for humans stems from the mind being fully absorbed in an activity, a state of mind that University of Chicago psychologist Mihaly Csikszentmihalyi describes as one of flow.[32] Steve is collaborating with Gabrielle Starr, an English professor at New York University, to use brain imaging to probe the structure of one form of optimal experience: aesthetic experience. Although objects of art come to mind most readily, aesthetic value permeates a great deal of what we do, from the sorts of objects we surround ourselves with to the projects we engage in. From the hiker who finds beauty in a rugged landscape to the mathematician who finds it in contemplating an esoteric mathematical object, our fulfillment as human beings depends on experiences of beauty and other aesthetic emotions. Steve and Gabrielle believe that the human desire to experience beauty stems from brain systems that give rise to aesthetic motivation, spurring us to engage in activities that afford experiences of aesthetic value. If we can better understand the nature of this experience, then we could apply these lessons to better design daily life. For this reason, Steve is also collaborating with Michael Dobry, codirector of the graduate industrial design program at the Art Center College of Design in Pasadena. The long-term goal of this collaboration is to use our emerging knowledge of the neural basis of aesthetic experience to guide the design of both objects and experiences, perhaps one day helping to better align the design of life with the needs created by our biology.

Not all attempts to rebuild engagement have to be so self-consciously based on lessons from the brain. If the pull of our biology is strong, then we should be able to see signs of its influence in other social trends. Consider the return to more traditional social modes—the return to ethnic identities, devolution, and the proliferation of emerging states. According to one explanation, corporations, the market, and technology all conspire to diffuse economic power in the world economy, increasingly marginalizing the nation-state, the achievement of modernity.[33] As the barriers to associational life disappear, new forms of associational life spring up from the pull of biology to participate.

In the United States, a growing distrust of government, a sense of disenfranchisement, and a realization that the federal government is the wrong level for service delivery have all conspired to sow the seeds of a quiet, local revolution. As Christopher Gates, publisher of the *National Civic Review*, puts it, "Our democracy is quietly and cautiously being repaired and reborn in our nation's communities. It is here that government and citizens are on the front lines of creating a new compact."

Known as community building or "new paradigm" communities, this shift in how Americans go about the process of civic improvement has captured enormous interest.[34] Community building shifts the locus of innovation to the local community, typically the one containing five thousand to ten thousand people. There, neighbors learn to work together on concrete tasks that take advantage of a new self-awareness of their collective and individual assets. Unlike traditional programs that are mandated from the top down, this process of community building has a wider set of objectives, which aim to build the friendships, mutual trust, and institutions that form the *social capital* essential for prospering communities. None of this is to say that such efforts are magical solutions or don't bring their own challenges; however, we believe the scale at which they operate resonates deeply with the human core, which was neglected by social policies that were based on economies of scale.[35]

The need of communities to own the processes of civic betterment resonates with lessons from field investigations of public goods dilemmas. In chapter 7, Garrett Hardin's answer to the "tragedy of the commons" reinforced the image of rational individuals trapped in social dilemmas from which they cannot extract themselves without inducement or sanctions applied from the outside. Field research on commons problems reveals that individuals can learn how to devise well-tailored rules and norms of cooperation when they participate in the design of the institutions affecting them. Externally imposed solutions, through either government ownership or privatization, often exacerbate the problems rather than solve them. This field research urges us to rethink how we approach problems of common concern. The modern solution—the heavy hand of Hobbes's Leviathan—might lead to as many problems as it was intended to solve. If society does not come about against human nature but flows up from it, then solving problems of common concern may depend on fostering local associations built from social capital.[36]

None of this implies crude laissez-faire social policy. Although we are not the asocial beasts Hobbes thought we were, neither are we angels. As history so readily reveals, our capacity for cooperation for mutual benefit doesn't alone guarantee the formation of civil society, nor is the link between civil society and governance unproblematic.[37] Rather, as exquisitely tuned to our contexts, the civil society necessary for human fulfillment requires a facilitating culture.[38] Consider the illustrative example of Russia. After the collapse of the Soviet Union, there was a lack of mediating mechanisms and institutions that could build and sustain a civil society. As a result, black and gray markets became major components of the economy and violent crime rates exploded.

Renewing Civil Society

Although the root causes of declines in social capital are complex, Robert Putnam's suggestion that television viewing is the culprit behind declining social capital hints that other emerging technologies, particularly the rise of the Internet, might also have dire effects on our happiness. Some technological pundits such as Howard Rheingold see in the Internet the promise of virtual communities that will free us from the limitations of geography.[39] According to them, we are moving toward a new, placeless society in which we will live in a disembodied existence over high-speed communication lines. Although a matter of controversy, initial studies of the effects of Internet use on well-being suggest that Internet use worsens our mental life. The first major study, by Carnegie Mellon's Robert Kraut and his colleagues, found that although participants used e-mail and Internet chat more than they used passive forms of online information-gathering like reading texts or watching videos, they reported a decline in interaction with family members and a reduction in their circles of friends that directly corresponded to the amount of time they spent online. They also reported increases in depression and loneliness. It will be interesting to see if further research confirms these findings. A recent study—the first assessment of the social consequences of Internet use based on a large, representative sample of American households including both Internet users and nonusers—finds a similar pattern. According to Norman Nie, the principal investigator of the study, "the Internet could be the ultimate isolating technology that further reduces our participation in communities even more than did automobiles and television before it."

These findings, although preliminary, suggest that the Internet might be yet another social parasite: It mimics the essential character-

istics of the social interactions your brain craves but lacks the necessary features to make them rewarding. Consider Internet communities, held together mainly by electronic chatting. Interactions in these
communities resembles to some degree those in real communities, yet
interactions in a physical community are based on the knowledge that
you have an enduring identity that holds you accountable for your actions over the long term. Your ability to engage in long-term reciprocal
interactions, from signing a mortgage to marriage, depends on the ability to sustain your identity across time. Internet communities lack this
constraint on interaction. Electronic commerce works only because it
rides on top of real-world social capital, although fraud is widespread
in some online auction sites because real community is hard to build
online. In fact, virtual communities are touted as places where people
are free to take on new, fluid identities, so-called onscreen avatars.
This precludes the construction of stable communities, which depends on long-term reciprocation. Perhaps these obstacles will be
overcome in the future, but the lack of face-to-face communication—
which plays an important role in facilitating group cooperation—suggests that virtual communities will remain too fluid, disembodied, and
transitory to build the relationships that generate social capital.

We wondered whether the popular idea that Internet communities
are somehow supposed to mimic real communities isn't a wrong way to
think about the use of the Internet. The communities you live in are
communities of place, physical living spaces where you spend your
days rooted in enduring face-to-face interactions. Virtual communities
appear to be more like communities built around processes instead of
places: associations tied together by common purpose and economic
interdependence rather than physical proximity. For example, members of the Sierra Club are tied together by common purpose, although
they don't exist as a physical community and have never gathered together as a single community. It seems to us that the communities
being developed on the Internet shouldn't try to emulate or substitute
for communities of place. The real promise of the Internet is that it
provides the tools to change how we structure daily life by reintegrating life's processes—work, education, recreation—back into local

communities. The growing postwar disillusionment was rooted in part in the fragmentation of modern life processes: Where we lived, where we went to school and work, and how we spent our leisure time grew increasingly distant from each other. The power of communication technologies, such as the Internet, lies in their capacity to perform many activities regardless of a specific physical location, freeing us to rethink how we should physically structure everyday life.

Consider how many pundits extol the coming of the electronic cottage. The idea is that if work no longer needs to be done at a specific location, then a worker can remain within his or her private residence. This seems like a surefire recipe for social isolation and depression. A more thoughtful response would involve asking the following question: If new technologies free some elements of work from a particular location, is there a way to place them in a context that will further human fulfillment?

Creating community-based work centers, where individuals spend a portion of their work time, could facilitate reintegrating work with other life processes. An influential group of architects and city planners known as the New Urbanists is designing communities with such work centers, which can also combine public spaces to serve as a focal point for the community.[40] While such developments are only in the early stages, they suggest that thoughtful use of new technologies can help root life in the local engagement of our human core desires. Thus, new technologies might really herald an emerging global village, but life will never be rooted in it. It is the wrong scale for primary human associations, and to be lulled by its promise is to fall into the technocratic trap that created much of the twentieth century's dehumanizing technology. Instead, if technology is to be aligned with human needs, it will do so by making life resemble a set of nested Russian dolls: multiple scales of life whose core is a reinvented local level, the community of place.

A Question of Culture

Rethinking the design of our living spaces and using emerging technologies intelligently are important steps toward rebuilding engagement. Yet there is perhaps a more central—and potentially more problematic—goal lying ahead. As creatures of meaning, we need our culture to provide the fabric from which we create a coherent self-conception to confer significance to our existence. Still, the great majority of people do not look to the culture of modernity for meaning. Rather, they look to religion for significance and answers to the problems of existence, both spiritual and practical. No doubt many people find religious answers to spiritual issues enormously helpful, yet, as Vernon Reynolds and Ralph Tanner explore, religions also provide ways of coping with the everyday realities of life: how to deal with one's own feelings and the actions of others, how to cope with the grief caused by death and natural disasters, and even how to regulate sexual behavior and structure family life.[41] In providing a prescribed user's guide to life, religions offer their followers images of both personal and group identity, and so provide an answer to the question Who am I?

In the twentieth century, the physical sciences created a powerful human epic, providing a naturalistic explanation of the origins of the universe from nothing, continuing to this day with such programs as NASA's Origins Program.[42] As the twenty-first century opens, a new chapter has begun, as we ourselves come into the purview of science. Indeed, cultural biology outlines an approach to answering who we are from a scientific perspective without dismissing the religious impulse as prescientific ignorance. That religion resurfaces whenever barriers to it are removed and that it flourishes in a time of unparalleled scientific knowledge testifies to the deep need within us for meaning.[43] These impulses no doubt helped our ancestors survive and for that reason remain with us today. Ultimately, brain science will understand the physical basis of these impulses, in the process deepening our appreciation for the needs our biology creates in us.[44] The great challenge ahead lies in whether we can create a sustaining culture—one that provides coherence to life—based on the advances of brain science.

If we are to use science's answers to who we are to create a culture of significance, a starting point might be one of biology's most surprising insights. At the foundation of a vision of who we are is a conception of what we all have in common, what unites us as human. This collective vision provides the basis for social and moral order by articulating common understandings about what is important in life and how to regulate life together. However, modern and postmodern culture has increasingly veered away from such commonality, often preferring foundations based on difference and predicated on the autonomous individual.[45]

Perhaps this is why groups in many parts of the world are returning to their ethnic identity, as it belongs to a premodern tradition that offers a coherent answer to who we are. In the United States, the rise of multiculturalism and identity politics has made us increasingly a nation of splitters, magnifying socially constructed differences that only leave room for private identities or special groups. This results in a conundrum. As Aristotle conjectured, our happiness depends in part on our social engagement in meaningful projects with others, yet contemporary culture appears incapable of articulating an inclusive view of who we.

Cultural biology points to a different conception of who we are that is rooted in our biological commonality and combines both construction of self and connection to other. Biology provides us with the capacity to construct flexible selves by endowing us with an internal guidance system that drives our construction of self into complex, multifaceted individuals. That we can become individuals who are distinct from others, however, in no way implies that we lack commonalities. Indeed, the journey to become individuals is driven by a biological core that creates needs universal to all. Primary among these is the need for social engagement and significance, mediated by the dopamine system and others we explored in chapter 7.[46]

It is an important, open question whether the new insights science offers into who we are can serve as the foundation both for a sustainable social order and for the rights we all deserve by virtue of the biology that makes us human. Ever since Darwin, the presumption was that our biology reinforces human differences among ourselves. Thus, the great irony of contemporary biology is that it provides the most compelling evidence against the divisive categories humans have used to rank others. With the first sequencing of the human genome now complete, the most surprising insight from biology is the overwhelming commonality of humans, and indeed of all of life. Consider that of the 3.2 billion base pairs in the human genome, individuals differ only by 2 million, and only a few thousand of these may account for the observed biological variability. All humans are nearly identical from a genetic standpoint. Devastating to the racial distinctions humans have made in the past, the biological unity of the human species and the common needs it creates provide a potential foundation for rethinking who we are, and perhaps contain the vision for thinking about our relation to one another.

Despite the titanic forces reshaping our world and the challenges they pose to fulfillment and happiness, we believe there is reason to be optimistic. Postmodern culture is now seen as a failure, in large measure because it offered nothing sustainable. In France, once the fount of postmodernism, the old thinkers have faded and a new generation is emerging, one that is focusing on a new civic republicanism rooted in the obligations and reciprocity of citizens to one another. In the United States, renewal at the grassroots level testifies to our biological drive to join others in enterprises with a common purpose and in projects larger than ourselves.

The oldest truly human search—the search for a social order in which we answer who we are—thus continues despite the barriers modern life sometimes puts in its way. The knowledge that brain science is beginning to provide signals a remarkable opportunity to deepen our appreciation for what makes us human. As we absorb these insights, they offer profound new lessons for structuring society in ways that can make life more fulfilling and significant, if we all enter the dialogue and provide wise counsel. For who we will be depends on the collective decisions we all make regarding how we will use the fruits of self-discovery.

Afterword

After September 11

When the plane moves and it's on its way . . . and you
come to the moment of close combat, then strike like
heroes who do not want to come back to Earth, say
ALLAHU AKBAR [God is great], because you will instill
terror in the infidel.

> —*Recovered instructions belonging to terrorists
> aboard three of the four airplanes hijacked
> during the September 11, 2001, attacks
> against the United States*

Mohamed Atta, the thirty-three-year-old
terrorist at the controls of the Boeing
767 that smashed into the North Tower
of the World Trade Center, and the ring-
leader who orchestrated the coordinated attack, had a
master's degree in architectural engineering from the
Technical University in Hamburg, Germany. His two sisters are university professors with Ph.D.'s. Before he
stepped onto American Airlines flight 11 and changed history, Atta had lived quietly in Florida for over a year. His
journey through life, from a middle-class upbringing in
Egypt, a professional education in Germany, conversion to

fundamentalist Muslim faith, recruitment into a terrorist cell, and the final unwinding of a diabolical suicide mission to destroy the infidel, is difficult for us to fathom.

The great paradox of being human—the human capacity for uncommon violence and uncommon compassion—was brought into focus on the morning of September 11, 2001. As the horrifying images of that day slowly recede and a flood of questions urgently in need of answers replaces the numbness of shock, the most pressing question is why—what could cause people to bring about such a senseless loss of life? Now in the midst of a war whose dimensions we don't fully know, addressing this question has an added urgency: Can understanding the root causes of militant Islamic terrorism provide any insight into possible paths to resolving this growing conflict?

An easy explanation is that the men who committed these atrocities were simply evil by nature. There is no doubt that the acts were deeply evil, but simply ascribing this to their nature leaves out the influences that guided them to their final end. The themes of *Liars, Lovers, and Heroes* suggest another route to understanding. From Philip Zimbardo's prison experiments in chapter 1 to Christopher Browning's historical documentation of the atrocities of Reserve Police Battalion 101 in chapter 8, we have seen that who we are is a complex blend of cultural and biological forces. Somehow the normal path to being human was corrupted by forces that created men who are willing to die in order to kill innocent people. What were these forces?

Throughout *Liars, Lovers, and Heroes,* we've highlighted how context shapes human behavior. A part of this context is the belief system of a culture. But do beliefs simply appear in a vacuum? Why do some beliefs take hold at certain times and places? To address these questions, it is crucial to ask what elements help shape a belief system. One element is the structural conditions around us. For example, Zimbardo's prison experiments illustrated that changing structural conditions—tossing average people into a prison environment—can have a profound influence on behavior by mediating new beliefs. Prisoner and guard alike developed new beliefs about one another, their roles, and their values and based what they did on those new beliefs. Hence, the prison environment caused profound behavioral changes by altering the belief systems of those living in that environment. So too conditions in everyday life profoundly affect what we believe.

Our starting point for attempting to make sense of the atrocities on September 11 and militant Islamic terrorism is the conditions behind the belief system that motivated those terrorists. Many people in the Islamic societies of the Middle East have experienced long-standing poverty with little prospect for improvement. The proportion of the population in these countries between the ages of fifteen and twenty-five exceeds 20 percent of the total. This youth bulge is a strong predictor of a country's violent crime levels and forms a ready pool of recruits for violent action.

However, Mohamed Atta was not an impoverished youth, but a well-educated and well-mannered adult, able to navigate the cultures in both the West and the East. Atta wasn't a suicide bomber recruited from a squalid camp, but a much more sophisticated and capable person caught in a collision between two worldviews. His conversion in Hamburg to Muslim fundamentalism and recruitment into a terrorist cell resemble more closely the dynamics of a cult, as we discussed in chapter 8. Atta, alienated from secular society, influenced by zealots who demonized the West and blamed the United States for their lot, was a living time bomb.

Religious ideologues like Osama bin Laden and Ayman al-Zawahiri, head of the radical Egyptian group Islamic Jihad, owe their rise to long-fermenting conditions in the Middle East. Since the 1960s, there has been a move from secular Islamic terrorism to religious-centered militant Islamic terrorism. Consider that in 1980 only two out of sixty-four militant Islamic groups were categorized as largely religious in motivation. By 1995, almost half of the identified groups were religiously motivated.[1] The Soviet invasion of Afghanistan and the subsequent anti-Soviet mujahideen war from 1979 to 1989 provided the circumstances for the rise of bin Laden's Al Qaeda. The Taliban militia in Afghanistan in 1994 provided logistical support and training facilities.

In 1998, bin Laden's Al Qaeda and al-Zawahiri's Islamic Jihad formed a coalition, the World Islamic Front for Fighting Crusaders and Jews.

The tendency to consider Islam as a homogeneous ideology obscures the cultural roots of the terrorist acts of September 11 and militant Islamic terrorism. Bin Laden, al-Zawahiri, and their network practice a conservative and austere brand of Islam known as Wahhabism, the state religion of Saudi Arabia (ten of the hijackers were from impoverished regions of Saudi Arabia).[2] The Wahhabi sect, an extreme minority in the Islamic world, propounds an extreme, antimodernist version of Islam that sees all popular Islam as idolatry and bans all Jews and Christians from the "holy land," which they define as the entire Arabian peninsula.

Bin Laden and his followers believe in a permanent jihad against infidels and other Muslims, especially the Shia sect.[3] His dream is to create a unified Islamic nation along Wahhabi lines, which for him would mean returning to how he imagines Islam was in the seventh century during the time of Mohammed. Taliban rule and society serve as his model, which in practice is a gruesomely authoritarian government.[4] In an escalating spiral of war and violence, open societies are at an unfair disadvantage.

The Wahhabi's antimodernist position has come into direct conflict with globalization, an Enlightenment idea, which is based on a mixture of liberal democracy and market capitalism. As we discussed in chapter 10, globalism and tribalism represent an intertwined dynamic of contrary movements. Globalization has spread a Westernizing influence, which is viewed as a threat in most Muslim nations and has led to strict traditionalism and even fundamentalism. The United States is used as a salient symbol of corrupting influences from the West, further polarizing the society and helping to solidify terrorist organizations such as Al Qaeda. The United States as a superpower provides a convenient Goliath to would-be Davids.

Given this potent mixture of structural and cultural influences, young men are easily recruited, drawn to the solidarity and sense of purpose these organizations offer. Once recruited, all the elements exist to create a culture of group violence, which we considered in chapter 8: cognitive fracture characterized by obsessive ideation, compulsive repetition, rapid desensitization to violence, blunting of emotional response, and hyperarousal, together acting like a contagion that spreads through the whole group. To this end, the rise of the Arab television network Al-Jazeera provides a convenient forum for bin Laden's recruitment efforts through anti-Western videos, which incorporate all the classic elements of a propaganda campaign.

Ideologues like bin Laden have existed throughout history, and the origins of their belief systems likely reflect their own idiosyncratic histories, though it requires a confluence of factors for them to rise to prominence. Hitler and the rise of Nazi Germany is an example of such a confluence. Bin Laden and his network, then, represent the fringes of human behavior that extreme structural and cultural influences can create. Centered on an extreme belief system and incubated in a culture of despair, this obsessive ideology has no doubt profoundly restructured the brain's value systems of those who adhere to it, to the point where they could believe that a religious deity would bless *fatwas* aimed at a civilian population. This dangerous brew of belief and structural conditions is certainly not unique in human history, though it seems particularly perverted from our perspective.

There are no quick fixes for long-standing poverty stemming from barely existent economies and authoritarian governments with a grip over hundreds of millions of people. The cultural problems are even more complex and represent systems of beliefs that go back more than a thousand years.[5] The great irony is that our ancestors created cultures as fonts of flexibility, yet today the apparently insatiable need to preserve culture is a fundamental stumbling block to removing the source of much hatred throughout the world.

This pessimistic view of recent violence in our midst should be balanced by other signs on the morning of September 11 and in the days following that are cause for optimism: the uncommon compassion of strangers helping strangers, putting themselves at risk to come to the aid of others. Many thousands of years ago in some distant land, our ancestors came together for mutual aid, propelled by the need to be in one another's company, creating the capacity for empathy that binds one to the other. This human urge remains and today brings us together in times of crisis. Although our future appears more uncertain than it has for many years, the human capacity for uncommon compassion is a source of hope for a better future.

October 21, 2001

notes

Chapter 1: Our Brains, Ourselves

1. Irene, a pseudonym, is based on an account relayed to Steve from a family friend and is not intended to be diagnostic.
2. Klaus's case and gourmand syndrome are described by Regard in M. Regard and T. Landis (1997). "Gourmad syndrome": Eating passion associated with right anterior lesions. *Neurology* 48:1185–90.
3. There is a long history of controversy surrounding claims of human uniqueness. Some of this debate is based on failing to make the distinction between differences across brains of distinct species, which is not controversial and forms the basis of comparative neuroanatomy, and inferences from differences to claims about comparative worth. The human brain differs from those of other species in many respects and confers capacities that may be unique to humans. However, other species have capacities that we lack that confer uniqueness on them and are remarkable achievements from an engineering perspective. Echolocation in bats is just one example of a capacity we lack. The more we learn about different brains, the more admiration we have for the natural processes of evolution that result in these complex systems. We leave it to the reader to decide whether making claims of worth from this diversity is productive.
4. We discuss this role of the state in solving problems of collective action in chapter 7. Freud's 1929 *Civilization and its discontents* (New York: W. W. Norton, 1989) is the classic twentieth-century account of the restraining role of civilization.

5.   For a historical account, see C. N. Degler (1991). *In search of human nature: The decline and revival of Darwinism in American social thought.* New York: Oxford University Press.

6.   The great anthropologist Franz Boas and his students, such as Ruth Benedict and Margaret Mead, advanced this view.

7.   E. Allen et al. (1975). Against "sociology." *New York Review of Books,* November 13, 182:184–86. For a sociological account of the debate surrounding sociobiology, see U. Segerstråle (2000). *Defenders of the truth: The battle for science in the sociobiology debate and beyond.* New York: Oxford University Press.

8.   Wilson's *Sociobiology* was primarily a synthesis of existing work on the social behavior of animals, and thus the intellectual framework of sociobiology was not novel. Others, such as David Lack, Robert Trivers, Richard Alexander, and William D. Hamilton, had already done seminal work in the area. It was Wilson's speculations regarding human society in his final chapter that made his work so controversial.

9.   Freud believed he was following in Darwin's footsteps; for the relation between the two, see F. Sulloway (1979). *Freud, biologist of the mind.* New York: Basic Books.

10.   R. Dawkins (1976). *The selfish gene.* New York: Oxford University Press. Taking a lesson from Wilson's reception, Dawkins avoided the term *sociology.*

11.   R. Dawkins (1995). *River out of Eden: A Darwinian view of life.* New York: Basic Books.

12.   The initial draft sequences of the human genome listed around thirty thousand genes, only twice as many genes as fruit flies have. See C. Venter et al. (2001). The sequence of the human genome. *Science* 291:1304–52; The Genome International Sequencing Consortium (2001). Initial sequencing and analysis of the human genome. *Nature* 409:806–921. For more information, see *Nature and Science*'s free access genomic websites at http://www.nature.com/genomics/human/ and http://www.sciencemag.org/feature/data/genomes/landmark.shl. However, only fifteen thousand genes were common to both lists. This suggests that the human genome has more than forty-five thousand genes; see J. B. Hogenesch, K. A. Ching, S. Batalov, A. I. Su, J. R. Walker, Y. Zhou, S. A. Kay, P. G. Schultz, and M. P. Cooke (2001). A comparison of the Celera and Ensembl predicted gene sets reveals little overlap in novel genes. *Cell* 106:413–15. It may take several more years to sort out the true number of genes in our genome.

13. This study suggests that humans and primates might not be suscepti-ble to the danger of clones growing abnormally large in the womb; see J. K. Killian, C. M. Nolan, A. A. Wylie, T. Li, T. H. Vu, A. R. Hoffman, and R. L. Jirtle (2001). Divergent evolution in M6P/IGF2R imprinting from the Jurassic to the Quaternary. *Human Molecular Genetics* 10:1721–28.

14. P. Baldi (2001). *The shattered self: The end of natural evolution.* Cam-bridge, Mass: Bradford Books.

15. Just as we no longer seriously consider prisons as places for rehabilita-tion, culpability need not be a requirement for sequestering individuals from society.

16. This poll was conducted by *U.S. News & World Report* in 1997 and was featured in "Born Bad?," the cover story of their April 21, 1997, issue. For the figures, see http://www.usnews.com/usnews/issue/970421/ nusurvey.htm.

17. N. Nicholson (1998). How hardwired is human behavior? *Harvard Business Review* 76:134–48.

18. R. Thornhill and C. T. Palmer (1999). *A natural history of rape.* Cam-bridge: MIT Press.

19. For an online overview of evolutionary psychology written by two of its founders, see J. Tooby and L. Cosmides, *Evolutionary Psychology: A Primer* (http://www.psych.ucsb.edu/research/cep/primer.html).

20. S. R. Quartz (in press). Toward a developmental evolutionary psychol-ogy: Genes, development, and the evolution of the human cognitive ar-chitecture. In *Evolutionary psychology: Alternative approaches,* edited by S. Scher and M. Rauscher. Boston: Kluwer.

21. J. Allman (1999). *Evolving brains.* New York: W. H. Freeman & Co.

22. This type of serial reaction time improvement depends on a subject's attending to the task, but not on explicit memory of the sequence or even the knowledge that there were repeating sequences; M. J. Nissen and P. Bullemer (1987). Attentional requirements of learning: Evi-dence from performance measures. *Cognitive Psychology* 19:1–32.

23. The logic of interpreting behavioral deficits from brain damage is, how-ever, not straightforward, since these might occur because there is an imbalance in some other part of the brain. If a radio makes a screech when a capacitor is removed, does this make the capacitor a screech suppressor?

24. Positron-emission tomography (PET) and functional magnetic reso-nance imaging (fMRI) indirectly measure brain activity through its ef-

fects on blood flow; M. I. Posner and M. E. Raichle (1994). *Images of mind*. New York: W. H. Freeman & Co.

25. G. S. Berns, J. D. Cohen, and M. A. Mintun (1997). Brain regions responsive to novelty in the absence of awareness. *Science* 276:1272–75.

Chapter 2: Making Connections

1. See http://socrates.berkeley.edu/~flymanmd/ for the Dickinson lab homepage.

2. Founded by the late Ed Posner, an information theorist at Caltech and the Jet Propulsion Laboratory, NIPS is an annual conference. Terry is the president of the NIPS Foundation, which oversees the NIPS conferences. The annual proceedings of the meeting are published by the MIT Press. For more information about these conferences, see http://www.cs.cmu.edu/Groups/NIPS.

3. An even more intimate gathering of scientists and engineers occurs at the Annual Teluride Workshop on Neuromorphic Engineering (http://www.ini.unizh.ch/telluride01), where researchers work together with computers and robots to test out their understanding of biological principles by building devices that work in the real world.

4. *Neural Computation,* founded by Terry in 1989 and published by the MIT Press, exemplifies the interdisciplinary nature of the NIPS conferences.

5. F. H. Crick (1979). Thinking about the brain. *Scientific American* 241(9):219–32.

6. Many terms in neuroanatomy are Latin; *substantia innominata* translates to "unnamed substance." It is a mysterious brain region containing neurons that modulate the function of the cerebral cortex. The pre-Botzinger complex, a small piece of the brain stem, is part of the respiratory oscillator and was named after a vineyard: in honor of the wine that was consumed while celebrating its discovery.

7. For example, what is called the "optic tectum" in the frog becomes the "superior colliculus" in the monkey. These are called homologous brain regions because they derive in the embryo from the same precursor cells and have similar patterns of connections with other brain regions.

8. The Sloan Foundation sponsored the meeting. The System Development Foundation subsequently sponsored much of the research by members of this group. The participants included David Rumelhart and Jay McClelland from psychology, Geoffrey Hinton, Teuvo Koho-

nen, and Jerry Feldman from computer engineering, and Stuart Geman
from statistics. Terry Sejnowski represented both physics and neuroscience. A book appeared based on that meeting: *Parallel models of associative memory.* G. E. Hinton and J. A. Anderson, eds. (1981).
Hillsdale, N.J.: Lawrence Erlbaum Associates.

9. Earlier work by pioneers such as Frank Rosenblatt and Norbert Wiener
in the 1950s had fallen into disfavor and been eclipsed by the promise
of an artificial intelligence based on logic and programming rather than
learning and dynamical systems.

10. For this reason, neural networks are often called "connectionist" architectures, and the scientists and engineers who design them are called
"connectionists." Some of the key connectionist discoveries were the
self-organizing map by Teuvo Kohonen, the attractor network of John
Hopfield, the Boltzmann machine by Geoffrey Hinton and Terrence
Sejnowski, and the backpropagation learning algorithm by David
Rumelhart, Geoffrey Hinton, and Ronald Williams. The results of
these discoveries were collected in a seminal two-volume book: D.
Rumelhart and J. McClelland, eds. (1986). *Parallel distributed processing.* Cambridge: MIT Press.

11. For a biography of Szilard see W. Lanouette and B. Silard (1992). *Genius in the shadows: A biography of Leo Szilard, the man behind the
bomb.* New York: Scribner.

12. The parts of the cerebral cortex that represent the outside world are the
best known, but there are other parts that represent the internal milieu,
such as the heart and the gut. We also use these circuits for thinking and
call them feelings. Antonio Damasio has written about these somatic
representations in his books: *Descartes' error* (New York: Putnam, 1994)
and *The Feeling of what happens* (New York: Harcourt Brace, 1999).

13. Steve Kuffler was a master at introducing new systems, or "prepara-
tions," into neurobiology and many are still used today. For example, he
was the first to record from ganglion cells in the retina of a living cat,
and ask what signals they were sending to the cat's brain about the visual world; S. K. Kuffler (1953). Discharge patterns and functional organization of mammalian retina. *Journal of Neurophysiology* 16:37–68.

14. At that time, positron-emission tomography (PET) was the main technique used for imaging the human brain; it relies on the injection of
radioactive tracers into the bloodstream and had a time resolution of
many minutes. Today, a technique called functional magnetic resonance imaging (fMRI) has made it possible to noninvasively study
changing activity in the human brain on a timescale of a few seconds.

15.   In Terry's laboratory, Charles Rosenberg created a neural network that converted English text to speech, called NETtalk, which caught the attention of the world. For a popular account of this era, see W. Allman (1989). *Apprentices of wonder: Inside the neural network revolution.* New York: Bantam.

16.   P. S. Eriksson, E. Perfilieva, T. Bjork-Eriksson, A. M. Alborn, C. Nordborg, D. A. Peterson, and F. H. Gage (1998). Neurogenesis in the adult human hippocampus. *Nature Medicine* 4:1313–17.

17.   H. van Praag, B. R. Christie, T. J. Sejnowski, and F. H. Gage (1999). Running enhances neurogenesis, learning, and long-term potentiation in mice. *Proceedings of the National Academy of Sciences USA* 96:13427–31.

18.   For a collection of nine essays regarding nonhuman primate behavior in relation to human evolution, see F. de Waal (2001). *Tree of origin: What primate behavior can tell us about human social evolution.* Cambridge: Harvard University Press.

19.   For some abilities, it appears that nonhuman primates are wading in this Rubicon.

20.   There are a number of issues in the naming of these two species. The bonobo *(Pan paniscus)* was once called the pygmy chimpanzee, but its size overlaps with the chimpanzee *(Pan troglodytes)*. Calling the bonobo a pygmy chimpanzee also forced the title "common chimpanzee" on its cousin, who thanks to human encroachment isn't so common anymore. The name bonobo itself is often attributed to a misspelling on a shipping crate from Bolobo, a town in Zaire.

21.   It appears more productive to consider the differences between human and nonhuman cognition as being on a continuum rather than sharply distinct. Many of these capacities have important antecedents in nonhuman primates. As Frans de Waal has studied, nonhuman primates appear capable of empathy and sympathy, two critical components of morality. They also are capable of complex communication and social transmission of behaviors, the foundation of culture. See F. de Waal (2001). *The ape and the sushi master: Cultural reflections of a primatologist.* New York: Basic Books.

22.   One of the most famous cases was Phineas Gage, who suffered a lesion of the orbitofrontal cortex when a rod was driven through the front of his brain while he was packing explosive charges; see A. Damasio (1994). *Descartes' error.* New York: Putnam.

23.   A description of the background to this announcement can be found in F. Crick (1994). *The astonishing hypothesis: The scientific search for the soul.* New York: Scribner, p. 265.

24. It is remarkable that in many patients suffering from severe depression, sleep deprivation will both increase the activity level in the anterior cingulate and relieve their depression. However, the depression and the depressed activity in the anterior cingulate return after a brief period of sleep; J. Wu, M. S. Buchsbaum, J. C. Gillin, C. Tang, S. Cadwell, M. Wiegand, A. Najafi, E. Klein, K. Hazen, and W. Bunney (1999). Prediction of antidepressant effects of sleep deprivation by metabolic rates in the ventral anterior cingulate and medial prefrontal cortex. *American Journal of Psychiatry* 156:1149–58.

25. D. A. Norman and T. Shallice (1986). Attention to action: Willed and automatic control of behavior. In *Consciousness and self-regulation*, edited by R. J. Davidson, G. E. Schwartz, and D. Shapiro. New York: Plenum Press, pp. 1–18.

26. A. Damasio (1994). *Descartes' error.* New York: Putnam.

27. K. Semendeferi, H. Damasio, R. Frank, and G. W. Van Hoesen (1997). The evolution of the frontal lobes: a volumetric analysis based on three-dimensional reconstruction of magnetic resonance scans of human and ape brains. *Journal of Human Evolution* 32:375–88.

28. K. Semendeferi, A. Schleicher, K. Zilles, E. Armstrong, and G. W. Van Hoesen (2001). Evolution of the hominoid prefrontal cortex: Imaging and quantitative analysis of area 10. *American Journal of Physical Anthropology* 114:224–41.

29. The increase in total dendritic length from primary cortical regions to area 10 is approximately 30 percent, and total spine number is about 60 percent greater in area 10 than in primary cortical regions; see B. Jacobs, M. Schall, M. Prather, E. Kapler, L. Driscoll, S. Baca, J. Jacobs, K. Ford, M. Wainwright, and M. Treml (2001). Regional dendritic and spine variation in human cerebral cortex: A quantitative Golgi study. *Cerebral Cortex* 11:558–71.

30. S. Harter (1999). *The construction of the self: A developmental perspective.* New York: Guilford Press.

31. The protracted nature of prefrontal cortex development has recently been examined with functional imaging; see J. N. Giedd, J. Blumenthal, N. O. Jeffries, F. X. Castellanos, H. Liu, A. Zijdenbos, T. Paus, A. C. Evans, and J. L. Rapoport (1999). Brain development during childhood and adolescence: A longitudinal MRI study. *Nature Neuroscience* 2:861–63.

Chapter 3: How to Make a Human

1.   This was the theme in Steven Spielberg's movie *A.I.*, released in 2001.

2.   The fastest supercomputers today can process around a teraOp, or around a million million operations per second. Each of the ten thousand synapses on the hundred billion neurons in the brain is activated on average once per second, which gives a minimum of a thousand million million operations per second. This could be much greater since the biochemical reactions inside each synapse also process information, though on a slow timescale.

3.   There are approximately ten times as many projections from the cortex back down to the thalamus, so the thalamus probably has more to do than simply relay sensory information to the cortex. During sleep, the feedback connections coordinate rhythmic activity between the thalamus and the cortex, which may be important for memory consolidation; see A. Destexhe and T. J. Sejnowski (2001). *Thalamocortical assemblies.* New York: Oxford University Press.

4.   This is an average percentage that ranges from 97.8 to 99 percent depending on the individuals being compared, with most of the variability coming from the differences between the genomes of individual chimpanzees. This figure includes a high proportion of noncoding DNA whose function is not known. In a comparison between twenty corresponding genes in chimpanzees and humans, there was a 99.3 percent agreement in their DNA sequences; see A. Varki (2001). A chimpanzee genome project is a biomedical imperative. *Genome Research* 10:1065–70. In addition to sequence differences between these species, there are also several large-scale chromosomal translocations, inversions, or transpositions, and regional deletions or duplications.

5.   There is growing evidence that small changes in the interactions between regulatory genes can be responsible for the large differences between species observed during development; see E. H. Davidson (2001). *Genomic regulatory systems: Development and evolution.* San Diego: Academic Press. Small changes in the interactions between genes responsible for brain development might similarly account for the differences between the bonobo brain and the human brain.

6.   E. O. Wilson (1998). *Consilience: The unity of knowledge.* New York: Alfred A. Knopf, p. 104. Although the long-distance pathfinding of axons does follow chemical guidance cues during brain development, the maintenance of fine precision in the connections depends on

activity-dependent mechanisms that are influenced by both internal and external stimuli.

7.   S. Pinker (1994). *The language instinct.* New York: William Morrow, p. 322.

8.   Despite the increasing evidence that Sperry's lock and key theory does not apply to the cerebral cortex, olfactory receptor neurons in the nose find their way to unique targets in the olfactory bulb even in the absence of electrical activity; see D. M. Lin, F. Wang, G. Lowe, G. H. Gold, R. Axel, J. Ngai, and L. Brunet (2000). Formation of precise connections in the olfactory bulb occurs in the absence of odorant-evoked neuronal activity. *Neuron* 26:69–80. The projection from the olfactory bulb to the olfactory cortex, by comparison, is diffuse and less precise.

9.   Building on these results, an array of impressive new experiments demonstrate that regions of the cortex can take on new information processing functions: L. von Melchner, S. L. Pallas, and M. Sur (2000). Visual behaviour mediated by retinal projections directed to the auditory pathway. *Nature* 404:871–76; J. Sharma, A. Angelucci, and M. Sur (2000). Induction of visual orientation modules in auditory cortex. *Nature* 404:841–47. In contrast to the malleability of the cortex, it is possible to transplant parts of the hypothalamus from a Japanese quail to a chick early in embryonic development, and the chick will grow up making the species-specific calls of a Japanese quail; see E. Balaban (1997). Changes in multiple brain regions underlie species differences in a complex, congenital behavior. *Proceedings of the National Academy of Sciences USA* 94:2001–2006.

10.   We don't want to give the impression that this is no more than just an interesting scientific oddity. The reason researchers were interested in transplanting fetal pig cells into rat brains was to see whether this might be a way to help repair damage to the human brain. In fact, fetal pig cells have been transplanted into the brains of patients suffering from Parkinson's disease, who show positive signs of recovery. Such transplants might be the next hope for a range of brain maladies.

11.   N. Sadato, A. Pascual-Leone, J. Grafman, V. Ibañez, M. P. Deiber, G. Dold, and M. Hallett (1996). Activation of the primary visual cortex by Braille reading in blind subjects. *Nature* 380:526–28. Transcranial magnetic stimulation studies demonstrated that this part of the brain is being used to process tactile information; see L. G. Cohen, P. Celnik, A. Pascual-Leone, B. Corwell, L. Falz, J. Dambrosia, M. Honda, N. Sadato, C. Gerloff, M. D. Catalá, and M. Hallett (1997). Functional

relevance of cross-modal plasticity in blind humans. *Nature* 389:180–83.

12. For a review of recent research on the molecular mechanisms of cortical patterning, see C. W. Ragsdale and E. A. Grove (2001). Patterning the mammalian cerebral cortex. *Current Opinion in Neurobiology* 11:50–58. For a recent review of the role of neural activity in shaping the barrel fields that were the focus of O'Leary and Schalgger's study, see R. S. Erzurmlu and P. C. Kind (2001). Neural activity: sculptor of "barrels" in the neocortex. *Trends in Neurosciences* 24:589– 95.

13. This surface is called the retina and contains an array of photo receptors, which convert light to electricity, and several layers of neurons, which process the electrical signals before they are sent through the optic nerve into the brain. The retina is a computational powerhouse and maybe one of the first parts of the brain that is understood at all levels, from its molecular composition to how it encodes the visual world; see J. E. Dowling (1987). *The retina: An approachable part of the brain.* Cambridge: Harvard University Press.

14. The study of training-induced plasticity was pioneered by Michael Merzenich; see D. V. Buonomano and M. M. Merzenich (1998). Cortical plasticity: From synapses to maps. *Annual Review of Neuroscience* 21:149–86.

15. The University of California at San Diego's Robert Katzman completed a study in Shanghai, China, showing that the more years of schooling, the older on average a person was before the onset of Alzheimer's disease. This study raises many issues, including the possibility that more education provides the brain with a greater reserve of neurons and synapses for later in life; see M. Zhang, R. Katzman, E. Yu, W. Liu, S. F. Xiao, and H. Yan (1998). A preliminary analysis of incidence of dementia in Shanghai, China. *Psychiatry and Clinical Neurosciences* 52:S291–94.

16. This is known as specific language impairment, which is believed to be an auditory problem. Hearing speech is a key to learning how to read. The average reading level of the deaf, for example, is a grade two level because of the lack of an auditory input. P. Tallal, S. L. Miller, G. Bedi, G. Byma, X. Wang, S. S. Nagarajan, C. Schreiner, W. M. Jenkins, and M. M. Merzenich (1996). Language comprehension in language-learning impaired children improved with acoustically modified speech. *Science* 271:81–84.

17. This is known as phonetic categorization. You never hear a sound that is halfway between two phonemes. Instead, through what is known as

the perceptual magnet effect, your brain pulls sounds into prototypes for phenomes, like a converter that decides what bin to place a signal into. You never hear the raw signal, but only the output of brain processors that clean things up for you; see P. K. Kuhl, K. A. Williams, F. Lacerda, K. N. Stevens, and B. Lindblom (1992). Linguistic experience alters phonetic perception in infants by 6 months of age. *Science* 255:606–608.

18. When babies are given a choice between parentese and adult talk, a kind of baby Nielson rating, they prefer parentese, even when the speaker is talking in a foreign language. It is also good therapy for stress in adults; see A. Gopnik, A. N. Meltzoff, and P. K. Kuhl (1999). *The scientist in the crib: Minds, brains, and how children learn.* New York: William Morrow.

19. J. McClelland has shown that with training on exaggerated contrasts, adult native Japanese speakers can be taught to distinguish between the *R* and *L* sounds; see J. L. McClelland (2001). Failures to learn and their remediation: A competitive, Hebbian approach. In *Mechanisms of cognitive development: Behavioral and neural perspectives,* edited by J. L. McClelland and R. S. Siegler. Hillsdale, N.J.: Lawrence Erlbaum Associates, pp. 97–121.

20. Based on the MacArthur Foundation Research Network on successful aging, an excellent reference to successful aging is J. W. Rowe and R. L. Kahn (1998). *Successful aging.* New York: Pantheon Books.

21. This intuitive way of thinking about genes is helpful, but it is important to note that it is an oversimplification. For a more complete account, see B. Lewin (1999). *Genes VII.* New York: Oxford University Press.

22. C. B. Ruff, E. Trinkaus, and T. W. Holliday (1997). Body mass and encephalization in Pleistocene Homo. *Nature* 387:173–76.

23. Although there is an overall increase of the brain size with the body size, the cerebral cortex in humans is greatly expanded compared with that in other primates, especially the white matter containing the bundles of fibers that connect neurons located in distant parts of the cortex; see J. M. Allman (1999). *Evolving brains.* New York: W. H. Freeman & Co.; K. Zhang and T. J. Sejnowski (2000). A universal scaling law between gray matter and white matter of cerebral cortex. *Proceedings of the National Academy of Sciences USA* 97:5621–26.

24. Known as selectionism, this was an ingenious way to explain complex brain structures without precisely wiring them. Recent evidence for the birth of new neurons and the growth of new synapses and dendrites in adults does not support selectionism; see S. R. Quartz and T. J. Se-

jnowski (1997). The neural basis of cognitive development: A constructivist manifesto. *Behavioral and Brain Sciences* 20:537–96; S. R. Quartz (1999). The constructivist brain. *Trends in Cognitive Sciences* 3(2):48–57.

25. Brain imaging is proving extremely powerful in supporting this human brain development; see J. N. Giedd, J. Blumenthal, N. O. Jeffries, F. X. Castellanos, H. Liu, A. Zijdenbos, T. Paus, A. C. Evans, and J. L. Rapoport (1999). Brain development during childhood and adolescence. A longitudinal MRI study. *Nature Neuroscience* 2:861–63; P. M. Thompson, J. N. Giedd, R. P. Woods, D. MacDonald, A. C. Evans, and A. W. Toga (2000). Growth patterns in the developing brain detected by using continuum mechanical tensor maps. *Nature* 404:109–93.

26. Known as concurrent synaptogenesis, this view was based on painstaking analysis of electron micrographs of primate brains; see P. Rakic, J. P. Bourgeois, M. F. Eckenhoff, N. Zecevic, and P. S. Goldman-Rakic (1986). Concurrent overproduction of synapses in diverse regions of the primate cerebral cortex. *Science* 232:232–35.

27. For a review, see J. M. Fuster (2001). The prefrontal cortex—an update. *Neuron* 30:319–33.

28. J. P. Schade and W. B. von Groenigan (1961). Structural organization of the human cerebral cortex: I. Maturation of the middle frontal gyrus. *Acta Anatomica* 47:72–111.

29. J. N. Giedd, J. Blumenthal, N. O. Jeffries, F. X. Castellanos, H. Liu, A. Zijdenbos, T. Paus, A. C. Evans, and J. L. Rapoport (1999). Brain development during childhood and adolescence: A longitudinal MRI study. *Nature Neuroscience* 2:861–63. Many studies find some decrease in brain sizes during adolescence, but it is important to note that this decrease occurs after most key events of cognitive development, so it does not support a selectionist view.

30. Evolutionary psychologists try to get around this by suggesting that postnatal development is rapid and less extensive. See the exchange between Steven Quartz and Steven Pinker at http://www.edge.org/documents/archive/edge19.html.

31. For a review, see S. R. Quartz (1999). The constructivist brain. *Trends in Cognitive Sciences* 3(2):48–57.

32. For a review, see S. R. Quartz (in press). Learning and brain development: A neural constructivist perspective. In *Connectionist models of development,* edited by P. Quinlan. New York: Psychology Press.

33. For a review, see J. M. Fuster (2001). The prefrontal cortex—an update. *Neuron* 30:319–33.

34. It is also important to note that higher cortical regions provide extensive feedback to lower ones, so that information flow goes both ways.

35. This doesn't mean we're jumping into the extreme of a blank slate. The brain provides many constraints, including subcortical organization, a basic cortical structure Dennis O'Leary terms the protocortex, and many others; see S. R. Quartz (1999). The constructivist brain. *Trends in Cognitive Sciences* 3(2):48–57.

36. For a review, see F. Hirth and H. Reichert (1999). Conserved genetic programs in insect and mammalian brain development. *Bioessays* 21:684.

37. E. H. Davidson (2001). *Genomic regulatory systems: Development and evolution.* New York: Academic Press.

38. For a review of this field, see R. A. Raff (2000). Evo-devo: The evolution of a new discipline. *Nature Reviews Genetics* 1:74–79. For an attempt to integrate evolutionary developmental biology and developmental neuroscience into a developmental evolutionary psychology, see S. R. Quartz (in press). Toward a developmental evolutionary psychology: Genes, development, and the evolution of the human cognitive architecture. In *Evolutionary psychology: Alternative approaches,* edited by S. Scher and M. Rauscher. Boston: Kluwer.

39. Barbara Finlay and Richard Darlington, both of Cornell University, examined the size of 11 different brain structures across 131 mammalian species, including humans, and found that big brains and small brains have strikingly similar relative sizes of their parts, suggesting that as brains get larger they do so in a coordinated, systematic way. B. L. Finlay and R. B. Darlington (1995). Linked regularities in the development and evolution of mammalian brains. *Science* 268:1578–84. Despite these similarities, different regions of the brain scale differently with the size of the brain. For example, the cortical white matter (axons that join brain areas) increases even more rapidly than the gray matter (neurons, synapses, and local connections) with increase in brain size; see K. Zhang and T. J. Sejnowski (2000). A universal scaling law between gray matter and white matter of cerebral cortex. *Proceedings of the National Academy of Sciences USA* 97:5621–26. When the relative sizes of different brain areas in the forebrain—the most forward part of the brain containing the cerebral cortex—are compared, even more striking relationships are found. Although the relative sizes of brain areas are highly conserved over a wide range of brain sizes within a group of species that are evolutionarily related, the proportions are different in different groups. For example, among insectivores, such

as the echidna or spiny anteater, almost half of the forebrain is devoted to olfactory processing, but this proportion is less than 10 percent for primates. There is an especially dramatic increase in the relative size of the forebrain devoted to the cerebral cortex among different primates, ranging from 50 percent in lemurs to 81 percent in monkeys and 95 percent in humans. Interestingly, the proportion for monkeys with larger social group sizes is 88 percent, which may be a result of the greater need for structures underlying social cognition.

40. For a review, see B. L. Finlay, R. B. Darlington, and N. Nicastro (2001). Developmental structure in brain evolution. *Behavioral & Brain Sciences* 24:263–308.

41. D. R. Kornack and P. Rakic (1998). Changes in cell-cycle kinetics during the development and evolution of primate neocortex. *Proceedings of the National Academy of Sciences USA* 95:1242–46.

42. R. A. Barton and P. H. Harvey (2000). Mosaic evolution of brain structure in mammals. *Nature* 405:1055–58.

43. D. A. Clark, P. P. Mitra, and S. S. H. Wang (2001). Scalable architecture in mammalian brains. *Nature* 411:189–93.

44. Investigations into specific genetic mechanisms underlying brain development and brain evolution are under way, research that will be greatly advanced by the sequencing of the human genome. For a recent review, see S. W. Wilson and J. L. Rubenstein (2000). Induction and dorsoventral patterning of the telencephalon. *Neuron* 28:641–51.

45. Terry Deacon has also made this important point; see T. W. Deacon (1997). *The symbolic species: The co-evolution of language and the brain*. New York: W. W. Norton.

46. The consequences of learning for evolution are also far-reaching, as first explored by James Mark Baldwin in 1896. A new factor in evolution. *American Naturalist* 30:441–51. For a bibliography on the Baldwin effect, see http://ai.iit.nrc.ca/baldwin/bibliography.html.

47. M. Constantine-Paton and M. I. Law (1978). Eye-specific termination bands in tecta of three-eyed frogs. *Science* 202:639–41.

48. The idea of robust flexibility and evolvability is discussed at length in J. Gerhart and M. Kirschner (1998). *Cells, embryos, and evolution*. Malden, Mass: Blackwell Science.

49. Terry Deacon has also come to a similar conclusion, and Stephen Wolfram, studying cellular automata, has shown how relatively simple programs can create complex structures that resemble a wide range of biological patterns. See his forthcoming book, *A new kind of science* (2002), Champaign, Ill.: Wolfram Media Inc.

50. See M. Donald (2001). *A mind so rare: The evolution of human consciousness*. New York: W. W. Norton.

51. M. Tomasello (1999). *The cultural origins of human cognition*. Cambridge: Harvard University Press.

52. In the mature state, this is known as theory of mind, a topic that has received a great deal of attention in recent years; see S. Baron-Cohen, H. Tager-Flusberg, and D. J. Cohen (2000). *Understanding other minds: Perspectives from developmental cognitive neuroscience*. 2d ed. New York: Oxford University Press. Whether or not nonhuman primates possess theory of mind is a matter of intense controversy; see C. M. Heyes (1998). Theory of mind in nonhuman primates. *Behavioral & Brain Sciences* 21:101–34.

53. Cornell's Urie Brofenbrenner calls this an ecological view of human development.

54. This process is known as joint attention. The baby observes the caregiver attending the object while simultaneously engaging with it.

Chapter 4: Life on the Far Shore

1. This account is based on the eyewitness field observations of the anthropologist Walter Goldschmidt; see W. Goldschmidt (1990). *The human career: The self in a symbolic world*. Cambridge, Mass.: Blackwell.

2. Shame is the subject of M. Lewis (1992). *Shame: The exposed self*. New York: Free Press.

3. The relationship between culture and a moral order is essential; see R. Wuthrow (1989). *Meaning and moral order*. Berkeley: University of California Press.

4. For an excellent review of the debate regarding culture in apes, see W. C. McGrew (1998). Culture in nonhuman primates? *Annual Review of Anthropology* 27:301–28. For a clear summary of recent research, see A. Whiten and C. Boesch (2001). The cultures of chimpanzees. *Scientific American* 284:60–67.

5. We should caution that this is an empirical hypothesis open to future experimental work. Research concerning the minds of nonhuman animals, and human infants for that matter, is extremely complex because protocols have to be developed for making inferences regarding internal states of mind on the basis of behavior. The fundamental barriers to communication make these inferences problematic.

6.   For an extensive discussion of these cognitive differences in terms of representational styles, see T. W. Deacon (1997). *The symbolic species: The co-evolution of language and the brain*. New York: W. W. Norton.

7.   Founded by Ajit Varki and supported by the Mathers Foundation, this group of researchers holds focused workshops in La Jolla, California, at the Salk Institute. Some of its members include John Allman at Caltech, Jim Moore at UCSD, Philip Lieberman at Brown University, Bernard Wood at George Washington University, Thomas Insel at Emory University, Jared Diamond at UCLA, Ernst Mayr at Harvard University, Sarah Hrdy at UC Davis, Jean-Jacques Hublin at CNRS in Paris, Kurt Benirschke at UCSD, and Terry Deacon at Boston University.

8.   Living and extinct human beings and their near-human ancestors are called "hominids" and belong to the Hominidae family of primates. They are not to be confused with "hominoids," which belong to the Hominoidea family of primates and include apes and humans.

9.   As Jane Goodall has shown, chimpanzees can also have a strong reaction to death, as in the case of a young chimpanzee whose mother's death left him wailing at her side for days, until he himself eventually died.

10. This sixty-thousand-year period represents only twenty-four hundred generations, assuming twenty-five years for each generation. This is not enough generations to produce a major advance through the mutation and natural selection of genes. This is, however, more than enough time for cultural evolution to produce major advances in technology and social innovation.

11. R. Foley (1995). *Humans before humanity*. Cambridge, Mass.: Blackwell.

12. In this exponential expansion the size continually doubles. For example, since 1950 the number of transistors on digital computer chips has doubled every eighteen months and is now reaching one hundred million on the largest chips. In the past, all such expansions have reached natural limits and slowed. Humans have extended natural limits in many ways but are still subject to the constraints imposed by finite resources.

13. Although the link between towns and agriculture is a traditional one, recent evidence suggests that the two may not have been tightly correlated as first supposed; see H. Pringle (1998). The slow birth of agriculture. *Science* 282:1446.

14. M. Ridley (1994). *The red queen: Sex and the evolution of human nature*. New York: Penguin.

15.   For a discussion of the idea of a homicide module, see D. M. Buss
       (1999). *Evolutionary psychology: The new science of the mind.* Boston:
       Allyn and Bacon. For a discussion of the idea of rape as an evolution
       strategy, see R. Thornhill and C. Palmer (2000). *A natural history of
       rape: Biological basis of sexual coercion.* Cambridge: MIT Press.

16.   R. Wright (1994). *The moral animal.* New York: Pantheon, p. 27.

17.   There's a caveat to all this ape comparison. Making sweeping general-
       izations about either species' behavior ignores considerable variation.
       Just as humans don't behave the same everywhere, neither do these
       apes.

18.   The female chimpanzee, however, is not simply a passive victim in all
       this. She has her own mating strategies, which include trysts with
       chimpanzees from outside her community. See P. Gagneux, C. Boesch,
       and D. Woodruff (1998). Female reproductive strategies, paternity, and
       community structure in West African chimpanzee. *Animal Behaviour*
       57:19–32.

19.   Studies of primate social behavior continue to change at a fast pace,
       making it worthwhile to bear in mind two caveats: There is a growing
       appreciation for behavioral flexibility within many ape species, particu-
       larly as ecological settings vary. Also, the study of bonobos continues to
       be contentious, as bonobos tend to be especially shy of humans and
       difficult to observe in the wild. For this reason, many studies involve
       bonobos in captivity, thus raising the question of whether such ob-
       served behavior reflects behavior in the wild.

20.   F. de Waal and F. Lanting (1997). *Bonobo: The forgotten ape.* Berkeley:
       University of California Press, p. 32.

21.   C. B. Ruff, E. Trinkaus, and T. W. Holliday (1997). Body mass and en-
       cephalization in Pleistocene Homo. *Nature* 387:173–76.

22.   For an introduction to the principles of paleoclimatology, see R. S.
       Bradley (1999). *Paleoclimatology: Reconstructing climates of the Quater-
       nary.* 2d ed. San Diego: Academic Press.

23.   W. H. Calvin (1998). The great climate flip-flop. *Atlantic Monthly*
       281:47–60.

24.   Mirror neurons in the cerebral cortex were discovered by Giacomo Riz-
       zolatti and his colleagues at the Institute of Human Physiology in
       Parma, Italy. These neurons activate either when the subject is watch-
       ing someone perform an action or when the subject himself performs
       the action. Mirror neurons are found in the frontal cortex (or-
       bitofrontal), the parietal cortex (7b), and the temporal cortex (superior
       temporal sulcus). Their function suggests that this system of neurons

would be useful for imitative learning of complex actions and may play a role in simulating the actions of others, which is thought to be a crucial capacity underlying theory of mind; see G. Rizzolatti, L. Fogassi, and V. Gallese (2001). Neurophysiological mechanisms underlying the understanding and imitation of action. *Nature Reviews Neuroscience* 2:661–70; G. Rizzolatti and M. A. Arbib (1998). Language within our grasp. *Trends in Neurosciences* 21:188–94; M. Iacoboni, R. P. Woods, M. Brass, H. Bekkering, J. C. Mazziotta, and G. Rizzolatti (1999). Cortical mechanisms of human imitation. *Science* 286:2526–28.

25. Richard Dawkins and others have argued for a close analogy between cultural and biological evolution; see S. Blackmore (1999). *The meme machine.* New York: Oxford University Press. There are many reasons why this analogy is problematic; for an overview, see U. Kher (1999). Is the mind just a vehicle for virulent notions? *Time* 153:53.

26. See M. Donald (1991). *Origins of the modern mind: Three stages in the evolution of culture and cognition.* Cambridge: Harvard University Press; R. Potts (1996). *Humanity's descent: The consequences of ecological instability.* New York: William Morrow.

27. M. Tomasello (1999). *The cultural origins of human cognition.* Cambridge: Harvard University Press.

28. See K. Nelson (1996). *Language in cognitive development: The emergence of the mediated mind.* Cambridge, Mass.: Cambridge University Press; J. W. Astington and J. M. Jenkins (1999). A longitudinal study of the relation between language and theory-of-mind development. *Development Psychology* 35:1311–20.

29. The "pinning" ceremony of marine recruits, in which badges are pinned through their skin inducing severe pain, is just another example of the rituals people in our society go through. The controversy surrounding fraternity hazing is another example. One evolutionary explanation is that participation in these rituals is a demonstration of one's commitment to cooperate with a group; for discussion, see B. Low (2000). *Why sex matters: A Darwinian look at human behavior.* Princeton, N.J.: Princeton University Press.

Chapter 5: Between Thought and Feeling

1. We are glossing over the distinctions between temperament, mood, and emotion. Your temperament is the stable part of your personality, your at-

titude toward life; your mood is your disposition toward your current situation; your emotions are transient reactions to specific events, such as surprise, fear, disgust, anger, sadness, and happiness. The distinctions between temperament, mood, and emotion depend primarily on the duration of the feeling and the boundaries between them are not distinct.

2.   Gender differences in serotonin synthesis is investigated in S. Nishizawa, C. Benkelfat, S. N. Young, M. Leyton, S. Mzengeza, C. de Montigny, P. Blier, and M. Diksic (1997). Differences between males and females in rates of serotonin synthesis in human brain. *Proceedings of the National Academy of Sciences USA* 94:5308–13.

3.   Neurons in the ventral tegmental area contain dopamine and project profusely throughout the forebrain, with a strong projection to the prefrontal cortex.

4.   A. Diamond (1998). Evidence for the importance of dopamine for prefrontal cortex functions early in life. In *The prefrontal cortex: Executive and cognitive functions,* edited by E. Angela, C. Roberts, E. Trevor, W. Robbins, et al. New York: Oxford University Press, pp. 144–64.

5.   The increasing complexity does not arise by creating new brain areas, but by enhancing the function of those that already exist. This is a different view from that of the triune brain of Paul MacLean, who divided the brain into a reptilian core, an emotional limbic system, and a rational neocortex, which evolved from the inside out; see P. D. MacLean (1990). *The triune brain in evolution.* New York: Plenum Press. All of these brain regions evolved in parallel, and it is a mistake to think that one area is more "primitive" than the other, since every brain system has components from all of these levels, strongly interacting and evolving together. For a good overview of these systems and how they interact with each other, see L. W. Swanson (2000). What is the brain? *Trends in Neurosciences* 23:519–27.

6.   For a related view that draws on cost-benefit analysis as well as life history regulatory systems, see P. La Cerra and R. Bingham (1998). The adaptive nature of the human neurocognitive architecture: An alternative model. *Proceedings of the National Academy of Sciences USA* 95:11290–94.

7.   M. Klam (2001). Experiencing ecstasy. *New York Times Magazine,* January: 21.

8.   3, 4-methylenedioxymethamphetamine is a cousin of amphetamine, or "speed." Ecstasy also releases dopamine, another brain chemical that affects mood in ways that we will discuss later in the book.

9.   There are fourteen different types of serotonin receptors in the brain.
        When serotonin binds to one of these receptors, it affects the target neu-
        ron and can change its function in many ways that can last for many days.
        The persistent individual differences in initiating appropriate, effective
        aggression, engaging in affiliative behavior, and participating in coopera-
        tive alliances in vervet monkeys correlate strongly with the density of the
        serotonin transporter and the serotonin-2A receptors in the orbitofrontal
        cortex and amygdala. Not everyone has the same reaction to Ecstasy be-
        cause everyone has different levels of serotonin to start with and has dif-
        ferent reactions to prolonged Ecstasy use. Serotonin is removed from the
        synaptic region by a serotonin transporter (this process is blocked by
        Prozac). The transporter reduces the amount of serotonin bound to re-
        ceptors and lowers the activity caused by the release of serotonin. The
        gene for the serotonin transporter is located on chromosome 17. One of
        the variants of this gene in the human population produces more trans-
        porter genes. Those who have two copies of this variant gene have more
        anger, hostility, depression, and impulsiveness.

10.   Prozac, also called fluoxetine, increases the level of serotonin activity in
        the brain by blocking reuptake (reabsorption into the nerve terminal)
        after it is released.

11.   In 1999, Prozac was the fifth bestselling prescription drug, at $2.6 bil-
        lion. Three serotonin reuptake inhibitors, Prozac, Paxil, and Zoloft,
        were among the top ten bestselling drugs in 2000, with total sales of
        $7.5 billion.

12.   G. L. Brown and M. I. Linnoila (1990). CSF serotonin metabolite (5-
        HIAA) studies in depression, impulsivity, and violence. *Journal of Clin-
        ical Psychiatry* 51:31–41.

13.   F. Saudou, D. Ait Amara, A. Dierich, M. LeMeur, S. Ramboz, L. Segu,
        M. C. Buhot, and R. Hen (1994). Enhanced aggressive behavior in
        mice lacking 5-HT1B receptor. *Science* 265:1875–78.

14.   P. T. Mehlman, J. D. Higley, I. Faucher, A. A. Lilly, D. M. Taub, J. Vick-
        ers, S. J. Suomi, and M. Linnoila (1994). Low CSF 5-HIAA concen-
        trations and severe aggression and impaired impulse control in
        nonhuman primates. *American Journal of Psychiatry* 151:1485–91.

15.   B. A. Golomb (1998). Low cholesterol and violence: Is there a connec-
        tion? *Annals of Internal Medicine* 128:478–87.

16.   Barry Jacobs at Princeton University has documented a close link be-
        tween repetitive motion and activation of serotonin neurons in the dor-
        sal Raphe nucleus, located in the brain stem; see B. L. Jacobs and C.

A. Fornal (1997). Serotonin and motor activity. *Current Opinion in Neurobiology* 7:820–25.

17.   For a firsthand account from a psychiatrist, see K. Redfield Jamison (1995). *An unquiet mind.* New York: Alfred A. Knopf.

18.   J. A. Gray (1973). Causal theories of personality and how to test them. In *Multivariate analysis and psychological theory*, edited by J. R. Royce. New York: Academic Press.

19.   This Latin term translates to "compact part of the black substance" and the neurons in it contain a black pigment that is visible to the unaided eye. Damage to this structure leads to Parkinson's disease caused by loss of dopamine in the forebrain; this devastating illness reduces both the initiation and the accuracy of motor responses and affects thinking.

20.   We should note that the 80 percent comes from people's self-report of their happiness, although an observer might not consider them happy (the important point is that people tend to see themselves as happy); see D. G. Myers and E. Diener (1995). Who is happy? *Psychological Science* 6:10–19.

21.   The levels of glucocorticoids in the blood, an objective measure of stress, decreases with income levels.

22.   For example, John McCain, who spent six years as a prisoner of war in Vietnam, went on to a distinguished career in politics.

23.   Among developed countries, it appears that it is not the richest societies that have the best health, but those that have the smallest income differences between rich and poor; see R. Wilkinson (1997). *Unhealthy societies: The affliction of inequality.* London: Routledge.

24.   E. Diener and C. Diener (1996). Most people are happy. *Psychological Science* 7:181–85 (p. 184).

25.   The relationship between dopamine and extraversion is extensively examined in R. A. Depue and P. F. Collins (1999). Neurobiology of the structure of personality: Dopamine, facilitation of inventive motivation, and extraversion. *Behavioral & Brain Sciences* 22:491–569.

26.   Martin Hammer was working in the lab of Randolph Menzel, a bee neurobiologist who had earlier worked out the pathways in the bee brain for learning to associate odors with sugar reward; see M. Hammer and R. Menzel (1995). Learning and memory in the honeybee. *Journal of Neuroscience* 15:1617–30.

27.   P. R. Montague, P. Dayan, C Person, and T. J. Sejnowski (1995). Bee foraging in uncertain environments using predictive Hebbian learning. *Nature* 377:725–28.

28. R. S. Sutton and A. G. Barto (1998). *Reinforcement learning: An introduction.* Cambridge: MIT Press. The term *temporal difference* refers to a key element in the learning process in which the expected reward is subtracted from the actual reward and the difference is used to make changes in the system. Neural mechanisms that could implement TD learning, called temporally asymmetric Hebbian learning, have been found in the brain; see T. J. Sejnowski (1999). The book of Hebb. *Neuron* 24:773–76.

29. Recently, Greg Berns, Read Montague, and colleagues have investigated this question using functional brain imaging and have discovered that these structures operate in humans much as they do in bees; see G. S. Berns, S. M. McClure, G. Pagnoni, and P. R. Montague (2001). Predictability modulates human brain response to reward. *Journal of Neuroscience* 21:2793–98.

30. This argument was used by linguists such as Noam Chomsky at MIT to ridicule psychologists such as Harvard's B. F. Skinner, who studied forms of conditioning in animals.

31. It is estimated that there are around 10^{40} different board positions in chess, so that cataloging or searching through them all would take Deep Blue 10^{34} years, much longer than the age of the universe.

32. G. Tesauro and T. J. Sejnowski (1989). A parallel network that learns to play backgammon. *Artificial Intelligence Journal* 39:357–90.

33. G. Tesauro (1995). Temporal difference learning and TD-Gammon. *Communications of the ACM* 38:58–68.

34. Perhaps the most important secret about the atomic bomb was that it worked, something that was not known before billions of dollars were put into the Manhattan Project. If you know ahead of time that there is some way to get a complex system to work it is more likely you would be willing to make the investment.

35. They implanted electrodes into the medial forebrain bundle, which is made of axons from dopaminergic cells in the ventral tegmental area that project to the forebrain.

36. Cocaine blocks the reuptake system for dopamine and other biogenic amines such as noradrenaline and serotonin.

37. The brain basis of pleasure is still poorly understood, however.

38. In the original experiments, Schultz recorded from dopamine cells in the ventral tegmental area. He later found nearly identical responses from dopamine cells in the *substantia nigra pars compacta;* see W. Schultz, P. Dayan, and P. R. Montague (1997). A neural substrate of prediction and reward. *Science* 275:1593–99.

39. The link between the dopamine system and decisions involves the prefrontal cortex, which receives inputs from the dopamine cells in the ventral tegmental area; see J. M. Fuster (2001). The prefrontal cortex—an update: Time is of the essence. *Neuron* 30:319–33. Prefrontal neurons display persistent activity that often outlives the presence of the sensory stimulus, which may help us temporarily keep online information needed for problem solving. This is a form of working memory; see P. S. Goldman-Rakic (1996). Regional and cellular fractionation of working memory. *Proceedings of the National Academy of Sciences USA* 93:13473–80. Dopamine tends to enhance persistent activity in the prefrontal cortex. When too little dopamine is delivered to the prefrontal cortex, performance on working memory tasks deteriorates, and when dopamine levels are too high, which occurs in schizophrenics, the persistent activity once established makes it difficult to switch off. In the intermediate range of normal dopamine levels, dopamine inputs from the ventral tegmental area could help influence decisions by selectively maintaining in working memory the information that is most likely to lead to the attainment of rewards; see D. Durstewitz, J. K. Seamans, and T. J. Sejnowski (2000). Neurocomputational models of working memory. *Nature Neuroscience* 3:1184–91.

40. N. Chomsky (1959). Review of verbal behavior. *Language* 35:26–58. Chomsky later reviewed Skinner's *Beyond freedom and dignity;* see N. Chomsky (1971). The case against B. F. Skinner. *New York Review of Books* 30:18–24.

41. It is often noted that the female pelvis is at the extreme edge of design given the constraints of bipedalism.

42. Different parts of the hypothalamus control different drives and regulatory systems, such as hunger, thirst, salt balance, and blood pressure, as well as sex-specific behaviors.

43. A difference in a single DNA base pair is called single-nucleotide polymorphism (SNP). The D_2 and D_4 dopamine receptors may differ among individuals.

44. This may also explain the phenomenal diversity of fetishes that humans acquire.

45. V. Reynolds and R. Tanner (1995). *The social ecology of religion.* New York: Oxford University Press. For a cognitive perspective, see P. Boyer (2001). *Religion explained.* New York: Basic Books.

Chapter 6: Becoming You

1. Recently, the navy handed Miramar over to the marines.

2. Temperament is a subset of personality that refers to your basic orientation to emotion and arousal, your sensitivity to stimulation, and your attentional responses.

3. The experience in the womb can, however, have a profound influence on fetuses, ranging from maternal hormonal influences to competition with twins for fetal resources.

4. The central noradrenaline, serotonin, dopamine, and acetylcholine systems in the brain are called neuromodulatory because they have very little direct effect on short timescales, but can strongly influence the responses of neurons to other inputs that provide direct excitation or inhibition to the neuron. For example, acetylcholine makes cortical pyramidal neurons more excitable—after the activation of cholinergic receptors on the neuron, the same input will produce more action potentials.

5. M. L. Schneider, E. C. Roughton, A. J. Koehler, and G. R. Lubach (1999). Growth and development following prenatal stress exposure in primates: An examination of ontogenetic vulnerability. *Child Development* 70:263–74; M. L. Schneider, A. S. Clarke, G. W. Kraemer, E. C. Roughton, G. R. Lubach, S. Rimm-Kaufman, D. Schmidt, and M. Ebert (1998). Prenatal stress alters brain biogenic amine levels in primates. *Development and Psychopathology* 10:427–40.

6. In some ways, this view is a modern version of the classical temperaments, except that the Greeks thought they arose from organs such as the spleen and the heart rather than the brain and the neuromodulatory systems do not correspond directly to traits. The Greek theories remain with us in expressions such as "having a lot of heart" or "being gutsy."

7. The change of a single amino acid in a protein, which can be caused by a single mutation of a base pair, can alter its function. The different versions of genes in a population are called polymorphisms.

8. Junk DNA may have an important function during stress or for regulation that we do not yet understand, so it may be premature to discount it too soon. In particular, repetitive sequences, which make up 50 percent of the human genome, may have an important role in evolution; see W. H. Li, Z. Gu, H. Wang, and A. Nekrutenko (2001). Evolutionary analyses of the human genome. *Nature* 409:847–49.

9. M. H. Johnson and J. Morton (1991). *Biology and cognitive development: The case of face recognition*. Cambridge, Mass.: Blackwell.

10. Rod Usher (1996). The Netherlands: Not a low country anymore. *Time* 148:16 (October 14, 1996), 72.

11. This is accomplished by measuring neurotransmitter metabolites in their cerebrospinal fluid.

12. As enticing as these studies are, we should also bear in mind that, as Kagan points out, one key difference between monkeys and humans is that in humans it's not just the objective conditions that must be considered. It is also necessary to consider how children perceive their situation. Similar experiments on humans are not possible, but there are circumstances when observations may help establish which behaviors can be influenced by parents. For example, the relationship between genetic factors and parental influence on criminal behavior was studied by Sarnoff Mednick and his team at the University of Southern California. They found that there was a statistically significant correlation between adoptees and their biological parents for convictions of property crimes. However, this was not true for violent crimes; see S. A. Mednick, W. F. Gabrielli, Jr., and B. Hutchings (1984). Genetic influences in criminal convictions: Evidence from an adoption cohort. *Science* 224:891–94.

13. Despite the media gloss on twin studies, today's estimates put the heritability of personality traits at around 40 percent. This means that in a certain environment, genetic factors account for less than half of the variance of personality traits. This doesn't mean that genes cause or determine less than half of your personality, since heritability estimates are tied to a specific environment (change the environment, and you change the heritability estimate). But it does highlight that genes and environment are partners, not warring competitors, in constructing personality.

14. R. D. Lane, E. M. Reiman, B. Axelrod, L. S. Yun, A. Holmes, and G. E. Schwartz (1998). Neural correlates of levels of emotional awareness: Evidence of an interaction between emotion and attention in the anterior cingulated cortex. *Journal of Cognitive Neuroscience* 10:525–35.

15. E. A. Nimchinsky, E. Gilissen, J. M. Allman, D. P. Perl, J. M. Erwin, and P. R. Hof (1999). A neuronal morphologic type unique to humans and great apes. *Proceedings of the National Academy of Sciences USA* 96:5268–73.

16. John Allman, personal communication.

17. As we explore in chapter 8, Freud would later add a second basic instinct, aggression, as his view of human nature darkened. In essence, Freud moved away from Rousseau's idea that society corrupted to Hobbes's idea that society is a necessary coercive force.

18. There is good evidence that sustained stress and high levels of corti-costeroids, a hormone secreted in response to stress, can permanently damage the hippocampus. See R. M. Sapolsky (1992). *Stress, the aging brain, and the mechanisms of neuron death.* Cambridge: MIT Press.

19. F. Sulloway (1996). *Born to rebel.* New York: Pantheon.

20. Ibid., p. 86.

21. J. M. Digman (1990). Personality structure: Emergence of the five factor model. *Annual Review of Psychology* 41:417–40.

22. H. Hartshorne and M. A. May (1928). *Studies in deceit: Studies in the nature of character.* New York: Macmillan; W. Mischel (1968). *Personality and assessment.* New York: Wiley.

23. W. Mischel and Y. Shoda (1998). Reconciling processing dynamics and personality dispositions. *Annual Review of Psychology* 49:229–59; W. Mischel and Y. Shoda (1995). A cognitive-affective system theory of personality: Reconceptualizing situations, dispositions, dynamics, and invariance in personality structure. *Psychological Review* 102:246–69.

24. W. Mischel and Y. Shoda (1998). Reconciling processing dynamics and personality dispositions. *Annual Review of Psychology* 49:229–59 (p. 246).

25. This perspective is also playing an increasing role in thinking about the nature of developmental psychopathology; see W. T. Boyce et al. (1998). Social context in developmental psychopathology: Recommendations for future research from the MacArthur Network on Psychopathology and Development. *Development & Psychopathology* 10:143–64.

26. B. Levine, S. E. Black, R. Cabeza, M. Sinden, A. R. McIntosh, J. P. Toth, E. Tulving, and D. T. Stuss (1998). Episodic memory and the self in a case of isolated retrograde amnesia. *Brain* 121:1951–73.

27. For a discussion of the distinction between human imitative learning and emulation learning, see M. Tomasello (1999). The human adaptation for culture. *Annual Review of Anthropology* 28:509–30.

28. H. L. Gallagher, F. Happé, N. Brunswick, P. C. Fletcher, U. Frith, and C. D. Frith (2000). Reading the mind in cartoons and stories: An fMRI study of "theory of mind" in verbal and nonverbal tasks. *Neuropsychologia* 38:11–21; D. T. Stuss, G. G. Gallup, Jr., and M. P. Alexander (2001). The frontal lobes are necessary for "theory of mind." *Brain* 124:279–86.

29. P. A. Russell, J. A. Hosie, C. D. Gray, C. Scott, N. Hunter, J. S. Banks, and M. C. Macaulay (1998). The development of theory of mind in

deaf children. *Journal of Child Psychology and Psychiatry and Allied Disciplines* 39:903–1001.

30. As in, for example, Bob Dole's convention speech in which he chastised Hillary Clinton ("with all due respect, it takes a family . . .").

31. C. Taylor (1989). *Sources of the self: The making of the modern identity.* Cambridge: Harvard University Press.

32. Adele Diamond and Pat Goldman-Rakic have studied the development of the prefrontal cortex and the inhibition of behavior in infants; see A. Diamond and P. S. Goldman-Rakic (1989). Comparison of human infants and rhesus monkeys on Piaget's AB task: Evidence for dependence on dorsolateral prefrontal cortex. *Experimental Brain Research* 74:24–40.

33. C. Bergh, T. Eklund, P. Sodersten, and C. Nordin (1997). Altered dopamine function in pathological gambling. *Psychological Medicine* 27:473–75; I. Perez de Castro, A. Ibanez, P. Torres, J. Saiz-Ruiz, and J. Fernandez-Piqueras (1997). Genetic association study between pathological gambling and a functional DNA polymorphism at the D4 receptor gene. *Pharmacogenetics* 7:345–48.

34. *Diagnostic and statistical manual of mental disorders,* or *DSM-IV,* the psychiatric community's official catalog of disorders, does not list sex addiction, though some suggest that it is a more appropriate term than sexual compulsion, since a compulsion is an action done over and over without apparent pleasure. For a general discussion, see A. Verghese (1998). The pathology of sex: Why can't some people stop having it? *New Yorker* 74:42–47.

Chapter 7: Friend, Lover, Citizen

1. Societies differ in their sense of fairness. In the "ultimatum game," someone is given a week's wage and told that they need to offer a part of it to an anonymous person that they know. The person can either accept the offer or reject it, in which case no one gets anything. In most Western societies, the most common offer was 50 percent, but members of the Machigeunga in a remote part of the Amazon rain forest in Peru routinely offered 15 percent or less, and the most common offer of Lamelara whalers from Indonesia was 58 percent. The amount offered did not correlate with age, sex, level of education, or affluence. The size of the offer was instead related to the degree of cooperation and market integration in the society. This suggests that some aspects

of morality may depend on lifestyle; see J. Henrich, R. Boyd, S. Bowles, C. Camerer, E. Fehr, H. Gintis, and R. McElreath (2001). In search of *Homo economicus:* Behavioral experiments in 15 simple societies. *American Economic Review* 91:73–78.

2.   Although Hobbes's state of nature is often portrayed as purely intellectual speculation, he actually wondered whether such native people as those in America lived in a state of nature. Indeed, it was common at that time to portray the nature/nurture controversy as one between what is shaped by society and what is given by nature.

3.   We should note that the notion of the scientific revolution is not itself without controversy. For a contemporary historical account, see S. Shapin (1996). *The scientific revolution.* Chicago: University of Chicago Press.

4.   Whereas a normative theory of self-interest prescribes that everyone ought to maximize their self-interest, Hobbes's theory of self-interest was a descriptive one, that people are psychologically so constituted that they can act no other way. This is known as psychological egoism.

5.   G. W. Sheldon (1991). *The political philosophy of Thomas Jefferson.* Baltimore: Johns Hopkins University Press.

6.   A. Smith [1776]. 1994. *The wealth of nations: An inquiry into the nature and causes.* New York: Modern Library.

7.   We should note that in later work Buchanan acknowledged that institutions could structure patterns of behavior that were other-directed; see J. Buchanan (1986). Then and now, 1961–1986: From delusion to dystopia. Paper presented at the Institute for Human Studies. For a historical overview of self-interest, see J. Mansbridge (1990). The rise and fall of self-interest in the explanation of political life. In *Beyond self-interest,* edited by J. Mansbridge. Chicago: University of Chicago Press, pp. 3–24.

8.   The traditional divide between human morality and animal behavior is increasingly questionable, suggesting a continuum of behavior; see F. de Waal (1996). *Good natured: The origins of right and wrong in humans and other animals.* Cambridge: Harvard University Press.

9.   Recently, there have been attempts to resuscitate group selection, although it requires certain evolutionary scenarios that are hard to justify and social norms that are highly restrictive, such as those found in closed religious groups; see E. Sober and D. S. Wilson (1998). *Unto others: The evolution and psychology of unselfish behavior.* Cambridge: Harvard University Press.

10. The first use of game theory in evolutionary biology was R. C. Lewontin (1961). Evolution and the theory of games. *Journal of Theoretical Biology* 1:382–403. The notion of an evolutionary stable strategy along with an extensive treatment of evolutionary game theory stems from J. Maynard Smith and G. A. Price (1973). The logic of animal conflict. *Nature* 246:15–18.

11. In the terms of evolutionary game theory, altruism is not an evolutionary stable strategy—it can be invaded by other strategies.

12. Defecting (ratting) is a dominant strategy.

13. A wide variety of dilemma games have been studied, see P. Kollock (1998). Social dilemmas: The anatomy of cooperation. *Annual Review of Sociology* 24:183–214. In addition, we should note that social dilemmas are studied by economists, political scientists, and sociologists, each of whom has their specialized terminology.

14. G. Hardin (1968). The tragedy of the commons. *Science* 162:1243–48; Hardin's article is reprinted online at http://www.dieoff.org/page95.htm.

15. In his *Treatise of human nature* (1739) David Hume constructed a related parable.

16. See E. Ostrom (1998). A behavioral approach to the rational choice theory of collective action: Presidential address, American Political Science Association. *American Political Science Review* 92:1–21.

17. Standard rational choice models predict a cooperation level of 0 percent, but consistently replicated findings demonstrate that people are willing to initially contribute 40 to 60 percent of the money given to them to play these games; see G. Marwell and R. Ames (1979). Experiments on the provision of public goods I: Resources, interest, group size, and the free-rider problem. *American Journal of Sociology* 84:1335–60. For a review, see J. Ledyard (1995). Public goods: A survey of experimental research. In *Handbook of experimental economics,* edited by J. Kagel and A. Roth. Princeton, N.J.: Princeton University Press, pp. 111–94.

18. R. M. Dawes, J. McTavish, and H. Shaklee (1977). Behavior, communication, and assumptions about other people's behavior in a common dilemma situation. *Journal of Personality and Social Psychology* 35:1–11. For a comprehensive meta-analysis, see D. Sally (1995). Conservation and cooperation in social dilemmas. A meta-analysis of experiments from 1958 to 1992. *Rationality and Society* 7:58–92.

19. E. Ostrom (1998). A behavioral approach to the rational choice theory of collective action: Presidential address, American Political Science Association. *American Political Science Review* 92:1–21.

20.   The German philosopher Immanuel Kant based his theory of ethics on such a notion of duty in his *Critique of practical reason* (1778).

21.   If we were to ask economists about where our preferences come from, many would say that's not the business of economics. Typically, economics is only interested in the relationships among your preferences once you have them, such as if you prefer chocolate ice cream over vanilla and vanilla over Chunky Monkey, then you had better prefer chocolate over Chunky Monkey. Others, such as Nobel laureate Gary Becker, who do explore the origins of our preferences, tend to cite social conventions.

22.   The dopamine neuromodulatory system includes both innate preferences and values and, most important, constrains the learning of new preferences.

23.   R. Dawkins (1976). *The selfish gene.* New York: Oxford University Press.

24.   C. Darwin (1871). *The descent of man.* New York: Random House, p. 478. The online version is available at http://www.infidels.org/library/historical/charles_darwin/descent_of_man.

25.   For an intriguing discussion of these ideas, see J. M. Allman (1999). *Evolving brains.* New York: W. H. Freeman & Co.

26.   Family units are found in diverse primates, such as owl monkeys, marmosets, and siamangs.

27.   More correctly, we should say the reinvention of the extended family, since many primates live in extended families. Extended families share both food and information; sharing information through teaching may facilitate brain maturation.

28.   J. M. Allman (1999). *Evolving brains* New York: W. H. Freeman & Co., p. 203.

29.   For an extensive treatment of these brains systems, see J. Panksepp (1998). *Affective neuroscience: The foundations of human and animal emotions.* New York: Oxford University Press. Some have speculated that there may also be a serotonin connection between romantic and obsessive-compulsive disorder; see D. Marazziti, H. S. Akiskal, A. Rossi, and G. B. Cassano (1999). Alteration of the platelet serotonin transporter in romantic love. *Psychological Medicine* 29:741–45.

30.   Whether such insights from animal models are directly relevant to humans is controversial. They nonetheless give us insights into what is possible and help us to design experiments to study the biological basis of human sexuality.

31. In rats, this is the preoptic area while in humans it is the interstitial nu-
     clei of the anterior hypothalamus. See, for example, S. Levay (1993).
     The sexual brain. Cambridge: MIT Press.

32. For a review, see T. J. Shors (2000). The modulation of memory forma-
     tion by stressful experience and sex differences in the brain. In *Model
     systems and the neurobiology of associative learning*, edited by J. Stein-
     metz, M. Gluck, and P. Solomon. Hillsdale, N.J.: Lawrence Erlbaum
     Associates, pp. 331–60.

33. Research suggests that 35 percent of sperm is lost after intercourse,
     whereas this number declines to 30 percent when a female reaches or-
     gasm. For a critical evaluation, see A. Dixson (1998). *Primate sexuality*.
     New York: Oxford University Press.

34. Because the signs of female orgasm are less telling than with male or-
     gasm, the criteria of female orgasm are harder to resolve; see A. Troisi
     and M. Carosi (1998). Female orgasm rate increases with male domi-
     nance in Japanese macaques. *Animal Behviour* 56:1261–65.

35. W. Blaicher, D. Gruber, C. Bieglmayer, A. M. Blaicher, W. Knogler, and
     J. C. Huber (1999). The role of oxytocin in relation to female sexual
     arousal. *Gynecologic and Obstetric Investigation* 47:125–26.

36. This figure includes opportunity costs such as lost wages and college
     education at a state school; see P. Longman (1998). The cost of chil-
     dren. *U.S. News & World Report* 124:50–56.

37. C. Ferris (1996). The rage of innocents. *The Sciences* 36:22–26.

38. This is called classical conditioning and is one of the best-studied
     forms of learning in psychology; see W. Schultz and A. Dickinson
     (2000). Neuronal coding of prediction errors. *Annual Review of Neuro-
     science* 2:473–500.

39. Aristotle's claim that humans are political animals is commonly misin-
     terpreted; political derives from *polis*, or city. Aristotle was thus claim-
     ing that we are social animals.

40. Reported in J. Archer (1997). Why do people love their pets? *Evolution
     and Human Behavior* 18:237–60.

41. T. Kellner (1998). Fido's insurance. *Forbes* (November 20), p. 144.

42. Reported in J. Archer (1997). Why do people love their pets? *Evolution
     and Human Behavior* 18:237–60.

43. Ibid.

44. J. Allman, A. Rosin, R. Kumar, and A. Hasenstaub (1998). Parenting
     and survival in anthropoid primates: Caretakers live longer. *Proceedings
     of the National Academy of Sciences USA* 95:6866–69.

45. L. F. Berkman and S. L. Syme (1979). Social networks, host resistance, and mortality: A nine-year follow-up study of Alamdea County residents. *American Journal of Epidemiology* 109:186–204.

46. I. Kawachi, B. P. Kennedy, and K. Lochner (1997). Long live community: Social capital as public health. *American Prospect* 35:56–59.

47. R. Putnam (2000). *Bowling alone: The collapse and revival of American community.* New York: Simon & Schuster. There is a great deal of controversy over these claims of diminishing social participation and social capital; for an alternative view, see E. Ladd (1999). *The Ladd report.* New York: Free Press.

48. K. Uvnas-Moberg (1998). Oxytocin may mediate the benefits of positive social interaction and emotions. *Psychoneuroendocrinology* 23:819–35.

49. K. Uvnas-Moberg, E. Bjokstrand, V. Hillegaart, and S. Ahlenius (1999). Oxytocin as a possible mediator of SSRI-induced antidepressant effects. *Psychopharmacology* 142:95–101.

50. There is evidence that some sort of empathy may be present in both infants and nonhumans, though this is controversial; see the debate in *Scientific American* online at http://www.sciam.com/1998/1198intelligence/1198gallup.html.

51. See L. Gaertner, C. Sedikides, and K. Graetz (1999). In search of self-definition: Motivational primacy of the individual self, motivational primacy of the collective self, or contextual primacy? *Journal of Personality & Social Psychology* 76:5–18.

52. G. Lakoff and M. Johnson (1980). *Metaphors we live by.* Chicago: University of Chicago Press.

53. L. Gaertner, C. Sedikides, and K. Graetz (1999). In search of self-definition: Motivational primacy of the individual self, motivational primacy of the collective self, or contextual primacy? *Journal of Personality and Social Psychology* 76:5–18.

54. R. M. Dawes, A. van de Kragt, and J. M. Orbell (1990). Cooperation for the benefit of us—not me, or my conscience. In *Beyond self-interest,* edited by J. Mansbridge. Chicago: University of Chicago Press, pp. 97–110.

55. K. Harry (1992). The truth about Jonestown. *Psychology Today* (March): 62–67.

56. Ibid.

Chapter 8: The Killer Within

1. See, for example, *The State of the World Conflict Report* at The Carter Center: http://www.cartercenter.org/UPDATES/updates.html.

2. *Statistics on Violence,* U.S. Department of Justice, Bureau of Justice Statistics. As disturbing as these trends are, UCLA criminologist James Q. Wilson suggests that if today's emergency care quality was at its 1957 level, today's murder rate would be three times what it is.

3. See D. M. Buss (1999). *Evolutionary psychology: The new science of the mind.* Boston: Allyn and Bacon.

4. R. Thornhill and C. T. Palmer (1999). *A natural history of rape.* Cambridge: MIT Press.

5. D. M. Buss (1999) *Evolutionary psychology: The new science of the mind.* Boston: Allyn and Bacon.

6. We aren't doubting that the proportion of young males between the ages of fifteen and twenty-nine is one of the best predictors of violent aggression in a society; we are doubting the inference from that observation to an explanation based on evolved violent strategies.

7. R. E. Tremblay (2000). The development of aggressive behavior during childhood: What have we learned in the past century? *International Journal of Behavioral Development* 24:129–41.

8. D. Goleman (1996). *Emotional intelligence.* New York: Bantam.

9. See J. M. Allman (1999). *Evolving brains.* New York: W. H. Freeman & Co.

10. The image of apes succeeding through brute force is largely a mythical one. Social success among apes depends largely on coalition forming.

11. Among adult human and nonhuman primates, the influence of heritable variables on central serotonin turnover rates is less than 50 percent and environmental variables account for more than one-half of the variance; see J. D. Higley, S. J. Suomi, and M. Linnoila (1996). A nonhuman primate model of type II excessive alcohol consumption? Part 1: Low cerebrospinal fluid 5-hydroxyindoleacetic acid concentrations and diminished social competence correlate with excessive alcohol consumption. *Alcoholism, Clinical and Experimental Research* 20:629–42.

12. The specific form of alcoholism is known as early-onset alcoholism; see D. Goldman (1996). Why mice drink. *Nature Genetics* 13:137–38; K. P. Lesch, D. Bengel, A. Heils, S. Z. Sabol, B. D. Greenberg, S. Petri, J. Benjamin, C. R. Miller, D. H. Hamer, and D. L. Murphy (1996). Asso-

ciation of anxiety-related traits with a polymorphism in the serotonin transporter gene regulatory region. *Science* 274:1527–31.

13. J. D. Higley, S. J. Suomi, and M. Linnoila (1992). A longitudinal assessment of CSF monoamine metabolite and plasma cortisol concentrations in young rhesus monkeys. *Biological Psychiatry* 32:127–45.

14. Various dissociative disorders are still a matter of controversy; see H. G. Pope, Jr., P. S. Oliva, J. I. Hudson, J. A. Bodkin, and A. J. Gruber (1999). Attitudes toward DSM-IV dissociative disorders diagnoses among board-certified American psychiatrists. *American Journal of Psychiatry* 156:321–23.

15. A. Damasio (1994). *Descartes' error*. New York: Putnam, p. 209.

16. Ibid. p. 218.

17. S. W. Anderson, A. Bechara, H. Damasio, D. Tranel, and A. Domasio (1999). Impairment of social and moral behavior related to early damage in human prefrontal cortex. *Nature Neuroscience* 2:1032–37.

18. A. Raine, M. Buchsbaum, and L. LaCasse (1997). Brain abnormalities in murderers indicated by positron emission tomography. *Biological Psychiatry* 42:495–508.

19. H. L. Gallagher, F. Happe, N. Brunswick, P. C. Fletcher, U. Frith, and C. D. Frith (2000). Reading the mind in cartoons and stories: An fMRI study of "theory of mind" in verbal and nonverbal tasks. *Neuropsychologia* 38:11–21.

20. A. Raine, T. Lencz, S. Bihrle, L. LaCasse, and P. Colletti (2000). Reduced prefrontal gray matter volume and reduced autonomic activity in antisocial personality disorder. *Archives of General Psychiatry* 57:119–27.

21. L. Mealey (1995). The sociobiology of sociopathy: An integrated evolutionary model. *Behavioral and Brain Sciences* 18:523–99.

22. D. T. Lykken (1997). The American crime factory. *Psychological Inquiry* 8:261–70.

23. J. Gilligan (1997). *Voilence: Reflections on a national epidemic*. New York: Vintage Books.

24. F. Fukuyama (1992). *The end of history and the last man*. New York: Free Press.

25. C. Browning (1992). *Ordinary men: Reserve Police Battalion 101 and the final solution in Poland*. New York: HarperCollins.

26. Ibid., p. 170.

27. There is a great deal of controversy over whether contact between ethnic groups tends to increase or decrease the chance of conflict. See, for example, D. Dion (1997). Competition and ethnic conflict: Artifactual? *Journal of Conflict Resolution* 41:638–49.

28.   See the story of Bertie Felstead, the last known World War I survivor who played no-man's-land football. He died on July 22, 2001, at age 106. *The Economist,* August 2, 2001.

29.   D. Grossman (1995). *On killing: The psychological costs of learning to kill in war and society.* Boston: Little, Brown, p. 254.

30.   Ibid., p. 255.

31.   Ibid., pp. 304–305.

Chapter 9: Inside Intelligence

1.   For an analysis of the cognitive challenges of the knowledge society, see E. Hunt (1995). Will we be smart enough?: A cognitive analysis of the coming workforce. New York: Russell Sage Foundation.

2.   Binet's practice of subtracting mental age from physical age doesn't capture age-related differences very sharply. Stern suggested that mental age be divided by physical age and then multiplied by 100. This stretches differences more in accord with our intuition. For example, a three-year-old who scores at a five-year-old level would equal $5/3 \times 100 = 166$, while a nine-year-old who scores at an eleven-year-old level would equal $11/9 \times 100 = 122$. Most of us would agree that the three-year-old is further ahead, an intuition Stern's scale captures.

3.   For historical accounts, see S. J. Gould (1981). *The mismeasure of man.* New York: W. W. Norton; D. J. Kevles (1985). *In the name of eugenics.* New York: Alfred A. Knopf.

4.   K. Ludmerer (1972). *Genetics and American society: A historical appraisal.* Baltimore: Johns Hopkins University Press, p. 25.

5.   H. Goddard (1917). Mental tests and the immigrant. *Journal of Delinquency* 2:271.

6.   In 1908, Ford introduced the Model T, the first mass-produced automobile in the world and one of the greatest success stories in the annals of business. It would change the world and make mass production the centerpiece of work in the twentieth century. Ford's sales in 1908–9 were just 10,607. But through applying the methods of mass production, which reduced the price of the automobile from around $850 to $360, sales skyrocketed to 730,041 cars just eight years later. Almost unbelievably, during that time Ford's market share went from a mere 9.4 percent to 48 percent. In all, he sold more than fifteen million Model Ts over twenty years, converting the automobile from an exotic toy of the wealthy to an indispensable part of everyday life.

7. For a biography of Taylor, see R. Kanigel (1997). *The one best way: Frederick Winslow Taylor and the enigma of efficiency.* New York: Viking.

8. M. R. Warren (1912). Medieval methods for modern children. *Saturday Evening Post,* March 12.

9. F. Bobbitt (1913). *The supervision of city schools.* Chicago: University of Chicago Press, p. 15.

10. E. Cubberley (1909). *Changing conceptualizations of education.* Boston: Houghton Mifflin.

11. Ibid., p. 47.

12. D. Tyack (1974). *The one best system.* Cambridge: Harvard University Press, p. 208.

13. It is important to point out that the IQ test was also vaunted in some quarters as an instrument of social mobility, since it could in principle be used to put educational access on a meritocratic basis.

14. D. Tyack (1974). *The one best system.* Cambridge: Harvard University Press, p. 209.

15. L. Terman (1923). *Intelligence tests and school reorganization.* Yonkers-on-Hudson, N.Y.: World Book Company.

16. See J. Bernstein, E. C. McNichol, L. Mishel, and R. Zahradnik (2000). *Pulling apart: A state-by-state analysis of income trends.* Washington, D.C.: Economic Policy Institute and Center on Budget and Policy Priorities.

17. R. Herrnstein and C. Murray (1994). *The bell curve.* New York: Free Press, p. 523.

18. Ibid., p. 526.

19. Heritability is a technical term from population genetics referring to the percentage of the variation of a trait in a population that can be accounted for genetically. High heritability does not suggest that a trait can't be modified by different environments.

20. For a recent interpretation of IQ as the product of biological-environment feedback, see W. T. Dickens and J. R. Flynn (2001). Heritability estimates versus large environmental effects: The IQ paradox resolved. *Psychological Review* 108:346–69. They demonstrate how initially small genetic differences can be amplified through environmental feedback, leading to large ability differences.

21. See R. J. Sternberg, E. L. Grigorenko, and D. A. Bundy (2001). The predictive value of IQ. *Merrill-Palmer Quarterly* 47:1–41.

22. It would be surprising if IQ wasn't measuring *something* relevant to cognitive functioning, since IQ tests have constantly been retooled to better correlate with measures of success.

23. The position that mental processes are not isolated but are distributed socially is known as distributed cognition; see G. Salomon (1993). *Distributed cognitions: Psychological and educational considerations.* Cambridge: Cambridge University Press.

24. Merlin Donald, a psychologist at Queen's University in Canada, has written extensively on the relationship between the evolution of culture and of mind, and has pointed out the central role that externalizing thought likely played; see M. Donald (1991). *The origins of the modern mind.* Cambridge: Harvard University Press.

25. See A. Clark (1997). *Being there: Putting brain, body, and world together again.* Cambridge: MIT Press.

26. See D. Norman (1993). *Things that make us smart.* New York: Addison-Wesley.

27. For a review, see W. T. Greenough, J. E. Black, A. Klintsova, K. E. Bates, and I. J. Weiler (1999). Experience and plasticity in brain structure: Possible implications of basic research findings for developmental disorders. In *The changing nervous system: Neurobehavioral consequences of early brain disorders,* edited by S. H. Broman and J. M. Fletcher. New York: Oxford University Press, pp. 51–70.

28. Ironically, this possibility was alluded to by the authors of *The bell curve,* who stated, "because intelligence is less than completely heritable, we can assume that, some day, it will be possible to raise the intelligence of children through environmental interventions. But new knowledge is required. Scientific research is the only way to get it" (p. 414).

29. Frank Miele, interview of Robert Sternberg (1995). *Skeptic* 3:72–80. This interview is online at http://www.skeptic.com/03.3.fm-sternberg-interview.html.

30. For a discussion of lifelong learning, see R. Mundane and F. Levy (1996). *Teaching the new basic skills.* New York: Free Press.

31. See W. K. Schaie (1996). *Intellectual development in adulthood: The Seattle longitudinal study.* Cambridge: Cambridge University Press.

32. Ibid.

33. A. P. Shimamura, J. M. Berry, J. A. Mangels, and C. L. Rusting (1995). Memory and cognitive abilities in university professors: Evidence for successful aging. *Psychological Science* 6:271–77.

34. S. J. Buell and P. D. Coleman (1981). Quantitative evidence for selective dendritic growth in normal human aging but not in senile dementia. *Brain Research* 214:23–41. The cells examined were in the region of the parahippocampal gyrus.

35.   M. Snowdon (2002). *Aging with grace: What the nun study teaches us about leading longer, healthier, and more meaningful lives.* New York: Bantam Doubleday Dell.

36.   D. A. Evans (1990). Estimated prevalence of Alzheimer's disease in the United States. *The Milbank Quarterly* 68:267–89; D. Rice et al. (1993). The economic burden of Alzheimer's disease care. *Health Affairs* 12:164–76.

37.   R. J. Haier, B. V. Siegel, Jr., A. MacLachlan, E. Soderling, S. Lottenberg, and M. S. Buchsbaum (1992). Regional glucose metabolic changes after learning a complex visuospatial/motor task: A positron emission tomographic study. *Brain Research* 570:134–43.

38.   W. Schaie (1996). *Intellectual development in adulthood: The Seattle longitudinal study.* Cambridge: Cambridge University Press, p. 356.

39.   Reviewed in E. Tracy (1995). Running for brain power. *Current Health* 22:22–23.

40.   K. R. Isaacs, B. J. Anderson, A. A. Alcantara, J. E. Black, and W. T. Greenough (1992). Exercise and the brain: Angiogenesis in the adult rat cerebellum after vigorous physical activity and motor skill learning. *Journal of Cerebral Blood Flow and Metabolism* 12:110–19.

41.   H. van Praag, B. R. Christie, T. J. Sejnowski, and F. H. Gage (1999). Running enhances neurogenesis, learning and long-term potentiation in mice. *Proceedings of the National Academy of Sciences USA* 96:13427–31.

42.   In response, Gardner often acknowledges that different intelligences interact, but that admission diminishes the coherency of the proposal of multiple intelligences. In addition, there is a high degree of arbitrariness in defining what constitutes an intelligence.

43.   C. R. Jack, R. C. Petersen, P. C. O'Brien, and E. G. Tangalos (1992). MRI-based hippocampal volumetry in the diagnoses of Alzheimer's disease. *Neurology* 42:183–88.

44.   J. Golomb, M. J. de Leon, A. Kluger, A. E. George, C. Tarshill, and S. H. Ferris (1993). Hippocampal atrophy in normal aging: An association with recent memory. *Archives of Neurology* 50:967–73.

45.   See R. Shephard (1996). Habitual physical activity and academic performance. *Nutrition Reviews* 54:32–36.

Chapter 10: The Search for Happiness

1.   Sawyer's interview with Terry aired April 5, 1999.

2.   The computer system automatically categorizes faces according to the pattern of facial actions, which roughly corresponds to the forty-four different muscles of the face; for background, see P. Ekman and W. Friesen (1978). *Facial action coding system: A technique for the measurement of facial movement.* Palo Alto, Calif.: Consulting Psychologists Press.

3.   The time resolution of functional magnetic resonance imaging at present is one to two seconds, not fast enough to follow the speed of thought but good enough to track shifts in emotion. Electrical recordings of the brain from the scalp, called the electroencephalograph (EEG), can easily capture the moment-to-moment processing in the brain, but this technique has poor spatial resolution. When it becomes possible to apply them together, "mind reading" may become a reality.

4.   P. Ekman (1984). Expression and the nature of emotion. In *Approaches to emotion,* edited by K. Scherer and P. Ekman. Hillsdale, N.J.: Lawrence Erlbaum Associates, pp. 319–43. Although they are universal, the conditions when these expressions occur vary widely. In Japan, public expression of emotion is rare, but in some Latin countries it is the norm. Thus, these basic expressions form a human core and individuals in a society learn when it is appropriate to show a feeling.

5.   P. Ekman (1973). *Darwin and facial expression: A century of research in review.* New York: Academic Press.

6.   A bibliographic compendium and resources on happiness can be found at http://ww.eur.nl/fsw/research/happiness/index.htm.

7.   For an examination of religion from a cognitive and evolutionary perspective, see P. Boyer (2001). *Religion explained.* New York: Basic Books.

8.   In important ways, neurosociology has its roots in a tradition of anthropologists, sociologists, and psychoanalysts who explored the effects of culture on personality. Among these are D. Riesman (1950). *The lonely crowd.* New Haven: Yale University Press; E. Fromm (1955). *The sane society.* New York: Holt, Rinehart and Winston; C. Lasch (1979). *The culture of narcissism.* New York: W. W. Norton; R. Bellah, R. Madsen, W. Sullivan, and S. Tipton (1985). *Habits of the heart.* Berkeley: University of California Press. Erich Fromm, in particular, urged that his position "was neither a biological one nor a sociological one if that

would mean separating these two aspects from each other" (*The sane society*, pp. 13–14).

9. For a classic historical analysis of he relationship between twentieth-century culture, society, and capitalism, see D. Bell (1976). *The cultural contradictions of capitalism*. New York: Basic Books.

10. As Harvard sociologist Daniel Bell notes, "the social revolution in modern society came in the 1920s, when the rise of mass production and high consumption began to transform the life of the middle class. In effect, the Protestant ethic as a social reality and a life-style for the middle class was replaced by a materialistic hedonism" (*The cultural contradictions of capitalism*).

11. Ibid., p. 70.

12. We should also note that this economic expansion was confined to white America.

13. W. Whyte (1956). *The Organization Man*. New York: Simon and Schuster.

14. E. Mayo (1945). *The social problems of an industrial civilization*. New York: Ayer, p. 111.

15. James Kunstler has explored the far-reaching implications of suburban life; see J. Kunstler (1996). *Home from nowhere: Remaking our everyday world for the twenty-first century*. New York: Simon and Schuster.

16. For a critique of suburban design and life, see J. Kunstler, *Home from nowhere*.

17. The sociologist Herbert Gans argued that city dwellers create "urban villages" within urban areas to foster social connectedness, but the sprawl of suburban living makes this prohibitive.

18. The orangutan, a loner, is the exception.

19. R. Putnam (1995). Bowling alone: America's declining social capital. *Journal of Democracy* 6:65–78.

20. This is a complex debate, in part because it is difficult to determine an appropriate set of measures of social capital and civic engagement; see E. C. Ladd (1999). *The Ladd report*. New York: Free Press.

21. It is interesting to note that there was a pronounced drop in social capital a century ago, as industrialization was first causing a major shift in the design of life.

22. M. Sandel (1996). *Democracy's discontent*. Cambridge: Harvard University Press, p. 224.

23. The view of technology as the engine of globalization has been examined in such works as A. Toffler (1980). *The third wave*. New York: William Morrow; E. Dyson (1997). *Release 2.0*. New York: Broadway Books.

24.   W. Knoke (1996). *Bold new world.* New York: Kodansha International.

25.   L. Thurow (1996). *The future of capitalism.* New York: William Morrow; B. Barber (1995). *Jihad vs. McWorld.* New York: Times Books.

26.   Reflecting on this, Harvard sociologist Daniel Bell observed, "The return to kinship or primordial attachments as the basis of bonding is, in a sense, the return to premodern modes of social structure."

27.   The 1985 surprise bestseller *Habits of the Heart* provided compelling evidence for the American desire to rebuild community life.

28.   The maximum human group size of 150 is an arbitrary cutoff based on a smooth curve.

29.   B. Fagan (1999). *Flood, famines, and emperors.* New York: Basic Books.

30.   There are other factors as well, including the possibility of self-selection (people who are community-minded might choose to live there) and geographical isolation (the community is geographically defined within a canyon that has little immediate physical contact with other communities).

31.   See J. Pfeffer (1998). *The human equation.* Cambridge: Harvard Business School Press.

32.   M. Csikszentmihalyi (1990). *Flow: The psychology of optimal experience.* New York: Harper & Row.

33.   S. Strange (1996). *The Retreat of the State.* Cambridge: Cambridge University Press; see also D. Held (1996). *Democracy and the global order.* Palo Alto, Calif.: Stanford University Press.

34.   See, for example, The Urban Institute's "Community building comes of age" at www.cpn.org/sections/topics/community/civic_perspectives/ cb_coming_of_age1.html. The rise of the "therapeutic state," from the New Deal to the Great Society, with is takeover of many roles previously performed by families and community institutions, was a reaction to the new scales of modern life. But it engendered a federal role in service delivery that exacerbated as many problems as it helped. This, too, is an enormously complex issue; for example, many social programs in the 1960s were formulated around projections of economic growth that did not continue in the early 1970s. See T. Schafer and J. Faux (1996). *Reclaiming prosperity.* Washington, D.C.: Economic Policy Institute.

35.   Johns Hopkins economist Lester Salamon refers to this shift in thinking about how we solve problems of collective action as an "associated revolution," which, he suggests, "may prove to be as significant to the latter twentieth century as the rise of the nation-state was to the latter nineteenth."

36.   See E. Ostrom, J. Burger, C. B. Field, R. B. Norgaard, and D. Polican-
      sky (1999). Revisiting the commons: Local lessons, global challenges.
      Science 284:278–82.

37.   For a critical discussion of some of these issues, see M. Foley and B.
      Edwards (1996). The paradox of civil society. *Journal of Democracy*
      7:8–52.

38.   The role of culture in economic and political life remains a controver-
      sial issue to pursue; see L. E. Harrison and S. P. Huntington (2000).
      Culture matters: How values shape human progress. New York: Basic
      Books.

39.   H. Rheingold (2000). *The virtual community: Homesteading on the elec-
      tronic frontier*. Cambridge: MIT Press.

40.   P. Katz (1994). *The new urbanism: Toward an architecture of community*.
      New York: McGraw-Hill. See also the Congress for the New Urbanism
      website at www.cnu.org.

41.   V. Reynolds and R. Tanner (1995). *The social ecology of religion*. New
      York: Oxford University Press.

42.   See the Origins Program website at http://origins.jpl.nasa.gov/.

43.   A. B. Newberg, E. G. D'Aquili, V. Rause, and J. Cummings (2001).
      Why God won't go away: Brain science and the biology of belief. New
      York: Ballantine Books.

44.   P. Boyer (2001). *Religion explained*. New York: Basic Books.

45.   The sociologist Anthony Giddens remarks, "The self in modern society
      is frail, brittle, fractured, fragmented"; see A. Giddens (1991). *Moder-
      nity and self-identity: Self and society in late modern society*. Palo Alto,
      Calif.: Stanford University Press, p. 169.

46.   Larry Arnhart has identified twenty basic human needs created by our
      biology; see L. Arnhart (1998). *Darwinian natural right*. Albany: SUNY
      Press.

Afterword

1.   Data for this afterword were based on the State Department's report
      Patterns of global terrorism at http://www.state.gov/s/ct/rls/pgtrpt/2000/.

2.   Saudi religious instructors trained the Taliban, and the Saudi govern-
      ment officially supported the Taliban until a few years ago, when bin
      Ladin expressed his contempt for the new Saudi exception to banning
      Jews and Christians from the Arabian peninsula. Bin Laden was par-
      ticularly outraged by the permanent presence of U.S. troops in Saudi

Arabia following the Gulf War. As a result, bin Laden started issuing *fatwas* that considered the Saudi monarchy infidel; he was forced to flee to the Sudan and then to Afghanistan, where he has turned with a vengeance on those who helped create him and his fortune.

3. Bin Laden has focused his invective on the American presence in Saudi Arabia, American support for Israel, and American support for Egyptian and other Arab governments he and his followers regard as corrupt.

4. Since being accepted by the Taliban, bin Ladin has spent a great deal of time with Mullah Mohammed Omar, their absolute leader and self-proclaimed Commander of the Faithful, explaining to him his vision of Islam.

5. It is not difficult for Islamic fundamentalists to find support for terror-ism in the Koran: "And kill them wherever you find them, and drive them out from whence they drove you out such is the recompense of the unbelievers" [2.191]; "They desire that you should disbelieve as they have disbelieved, so that you might be [all] alike; therefore take not from among them friends until they fly [their homes] in Allah's way; but if they turn back, then seize them and kill them wherever you find them, and take not from among them a friend or a helper" [4.89].

index

foreign languages, 43
France, 277
free riders, 157
free will, 13–14, 29
Freud, Sigmund, 7, 11, 131–35, 149, 152, 193, 257, 263, 267–68, 309n
Fried, Itzhak, 206–7
Friedan, Betty, 260
friends, 176–78
Friends, 262
frogs, 25–26, 56–57, 165, 288n
Fromm, Erich, 1, 145, 323n–24n
Fukuyama, Francis, 204
functional magnetic resonance imaging (fMRI), 39, 287n, 289n, 323n

Gabrielli, John, 253
Gage, Fred, 27, 39, 242, 248
Gage, Phineas, 290n
Galápagos Islands, 68–69, 133–34
Galton, Francis, 221
Galton Society, 221
gambling, 146–47
game theory, 155–58, 313n
ganglia, 25–26
Gans, Herbert, 324n
Gardner, Howard, 249, 322n
Gates, Christopher, 270
Geertz, Clifford, 144
genes:
  chemical messenger proteins and, 44
  homeobox, 53
  human variation in, 277
  Mendel's resource on, 220–21
  precise programming seen in, 37, 39, 40, 50, 51
  regulation of, 44–46, 53, 292n
  role of, 44
  selfish, 11–12, 158, 159–60, 193
  temperaments and, 125–28
Genetics, 221
genomes:
  size of, 53
  see also human genome
Germany, Nazi, 205–6, 282
Gilligan, James, 191, 203
glaciers, 78–79
Glenn, John, 124, 127
globalization, 264, 281
glucocorticoids, 305n
glutamate, 33
Goddard, Henry, 222–23, 224
Goleman, Daniel, 194
Goodall, Jane, 191, 300n
Good Morning America, 252
gorillas, 64, 65, 74, 179
Gould, Stephen Jay, 9
gourmand syndrome, 3
Gray, Jeffrey, 98
Great Society, 264
Greene, Maurice, 235

Greenough, William, 248
Grossman, Dave, 211–12
group killing, 204–8
group life, 261, 265–66, 267–68
Gua (chimpanzee), 36–37

Hall, Rob, 216–17, 264
Hamlet (Shakespeare), 102
Hammer, Martin, 103
Hansen, Doug, 216
happiness, 98–100, 252–77
  civil society and, 261–62, 272–74
  evolution and, 253–54
  and rise of cities, 256–61
  social engagement and, 269–72
  television and, 262–63
  work and, 258–59
Harary, Keith, 187
Hardin, Garrett, 157, 271
Harris, Eric, 210–11, 213
Harris, Judith Rich, 133, 135, 143
Harvard Business Review, 14
Heaven's Gate, 186–87
Hebbian learning, 306n
Hereditary Genius (Galton), 221
heroin, 167
Hersch, Patricia, 214
heterochrony, 54
hierarchy of needs, 121
highways, 260
hippocampus, 42, 197, 242, 248, 249
Hitler, Adolf, 282
Hobbes, Thomas, 143, 149, 151–52, 160, 163, 178, 187, 205, 257, 267, 271, 309n, 312n
homeobox genes, 53
Homo erectus, 65
Homo habilis, 65, 162
Homo sapiens, archaic, 65
Hooking Up (Wolfe), 252
human genome:
  bioinformatics and, 234
  chimpanzee genome compared with, 36, 292n
  sequencing of, xiv, 12, 277, 286n
Human Genome Project, 12
humans:
  archaeological record of, 66–68
  cloning of, 13
  evolution of, 64–66, 161–62, 300n
  protracted development of, 46–52, 113, 131
  sexual dimorphism in, 161
  uniqueness of, 4, 285n
Huntington's disease, 18
Huxley, T. H., 150
hyenas, spotted, 11, 193
hypothalamus, 35, 114, 165, 166, 174, 293n, 307n

ideologies, 121
immigration, 222–23, 227